GONZO SCIENCE

THIS BOOK IS FOR OUR FATHER

GONZO SCIENCE
Anomalies, Heresies, and Conspiracies

Jim Richardson and Allen Richardson

PARAVIEW PRESS

New York

CONTENTS

PART 1:
Critical Thinking and Skepticism

CHAPTER 1

The Gonzo Manifesto

WE USE THE TERM "GONZO" IN THE SENSE THAT OUTLAW journalist Hunter S. Thompson used it. It connotes weirdness, excitement, and danger. Gonzo also refers to Thompson's guerrilla style as he conducted raids on objectivity. Thompson was trying to say: "Enough of this B.S., I'm tired of journalism pretending to be objective. The story is actually about the reporter." It is our observation that science is prone to the same argument. So-called scientific facts amount to the current beliefs of scientists.

Dominant paradigms collapse; that is what they do. The trick is to use the holes in the current theory to see what you might expect from the up-and-coming theory.

Thompson may be a stylistic influence, but T.S. Kuhn's *Structure of Scientific Revolutions* figures strongly here, of course, as does Schick and Vaughn's critical-thinking classic, *How To Think About Weird Things*, and the exhaustive scientific anomaly catalogs of William R. Corliss, which can be found at Sciencefrontiers.com.

What Is a Gonzo Scientist?

A gonzo scientist is under no obligation to be a trained scientist. As free agents gonzo scientists can avoid the pitfalls and influences of the culture of science. These include having to play ball with the academy, begging for funding from military-industrial sources, and submitting to the vice-like grip of peer-reviewed journals, which rigidly enforce the status quo. The gonzo scientist is a science writer, thinker, commentator, and/or critic. He or she studies and comments on the system, but is not of it. This follows "the world's greatest paleontologist" David Raup's observation that people outside of a discipline may have fresher perspectives than those stuck inside of it.

A gonzo scientist is a generalist. It is much lamented that science today is in a rut of specializations. Scientific disciplines have sprouted a thousand narrow sub-disciplines, each with its own jargon, culture, inherent biases, and blind spots. Scientists are by and large schooled up to the hilt in one of these particular specialties, knowing everything there is to know about a very narrow topic. The generalist, on the other hand, draws from wide experience in many different fields, and is in a better position to see where they overlap, interconnect, and inform each other.

A gonzo scientist is an anomalist. Whether there is an actual science of "anomalistics" is an open question. What we do know is that more often than not, instead of being brought to light, many scientific anomalies are either blatantly ignored or unceremoniously shunted off stage with an *ad hoc* explanation, an explanation whose reason for being is to make an anomaly go away, saving the dominant theory from the embarrassment of being unable to explain something. What seems clear to us is that if you collect the anomalies in any field, you are in effect assembling a negative image of the counter-theory to the dominant one.

A gonzo scientist is a critical thinker. We have no patience for those who engage in fuzzy thinking or anything but the most rigorous consideration of ideas. By critical thinking we refer to the collection of cognitive skills necessary to avoid logical pitfalls and to identify good ideas amid the bad ones. It is useful, therefore, to pay attention to the community of professional scientific skeptics, because they have knowledge of and promote the use of these skills. Whether they always apply them appropriately is another matter.

We are basically disaffected scientific skeptics. We grew tired of the way our fellow skeptics wrapped themselves in the language of critical thinking whilst ascribing logical fallacies to everyone else. In this sense we are "zetetic" skeptics: we doubt *all* claims—even those of the professional skeptics—and everything gets a fair hearing.

Who Are the Professional Skeptics?

The keepers of the scientific faith are the professional skeptics, represented by the unfortunately named organization CSICOP (the Committee for the Scientific Investigation of the Claims of the Paranormal) and other groups. They fancy themselves the most rational rationalists, the most mechanical mechanists, and the most reductive reductionists. Their mission is to save us from the kook, the crank, the crackpot, and the pseudoscientist. But in their zeal to defend the mundane, they've developed a pathological aversion to the anomalous.

Armed with a cognitive toolbox containing strict criteria of adequacy, and on the lookout for logical pitfalls, these grumpy skeptics go around tilting at any old windmill.

What Are the "Criteria of Adequacy"?

The criteria of adequacy are used to distinguish good ideas from bad. These criteria are testability, fruitfulness, scope, simplicity, and conservatism. This gets tricky right around the conservatism part.

The criterion of conservatism states that the best theory is the one that is most in agreement with well-established beliefs. So let's say you're looking for a theory that explains how the universe came to be and the principles that determine its organization. To apply conservatism, you ask a cosmologist or an astronomer about the most established beliefs of their profession. They will tell you that the universe began in a Big Bang and then organized itself through gravity.

But long-standing ideas in the field of plasma physics undermine this notion because plasma physicists do not believe in the Big Bang. The most conservative view within this community is that the universe is organized around electromagnetic principles, and that a Big Bang origin is an unnecessary hypothesis. The plasma physicists have conservatively applied their science and arrived at completely different conclusions from other scientific conservatives.

Conservatism is not only unreliable in this way, but also bears a close resemblance to the notorious logical fallacy known as the "appeal to tradition." You are guilty of committing the appeal to tradition when you argue that something must be the case because that is the way it has always been. This common pitfall is illogical. Notice that conservatism and the appeal to tradition are virtually indistinguishable. This amounts to nothing less than a fault line running through the scientific method.

The way to inoculate science against the appeal to tradition is to jettison conservatism from the criteria of adequacy. This would leave the other criteria—testability, fruitfulness, scope, and simplicity—intact.

A conservative science is blind to its own appeals to tradition, and dismissive of anomalies. But it is the anomalies in a theory which point the way to newer, better theories.

The skeptics guard our rear flank, intercepting the Bible-thumpers and neutralizing other nuisances like the holocaust deniers and the New Age UFO religions. But the criterion of conservatism, although useful, can hold people back from discovering new and exciting facts and connections. Conservatism—almost by definition—will never lead science forward. Eject it from the criteria of adequacy, and get someone down here who can explain all these anomalies!

Halfway between the hardest science and the most far-out speculation, we will take the best of both. Gonzo science combines the critical thinking of the skeptics with the fearlessness and progressive thinking of the iconoclast. It's time to get gonzo.

Recommended Reading: *How To Think About Weird Things* by Theodore Schick Jr. and Lewis Vaughn; *Science Frontiers: Some Anomalies and Curiosities of Nature* compiled by William R. Corliss; and see *Dark Matter, Missing Planets, and New Comets* by Tom Van Flandern for a discussion of "the unscientific method."

CHAPTER 2

Critical Thinking for Dummies

MICHAEL SHERMER'S BOOK *BORDERLANDS OF SCIENCE* introduces us to a system of categorization for scientific areas. Since Shermer is an influential skeptic, we feel it is important to scrutinize his methods of evaluating claims. The items in Shermer's "Boundary Detection Kit" include questions such as:

1. How reliable is the source of the claim?
2. Does this source often make similar claims?
3. Have the claims been verified by another source?
4. How does this fit in with what we know about the world and how it works?
5. Has anyone, including and especially the claimant, gone out of the way to disprove the claim, or has only confirmatory evidence been sought?
6. In the absence of clearly defined proof, does the preponderance of evidence converge to the claimants' conclusion, or a different one?
7. Is the claimant employing the accepted rules of reason and tools of research, or have these been abandoned in favor of others that lead to the desired conclusion?
8. Has the claimant provided a different explanation for the observed phenomena, or is it strictly a process of denying the existing explanation?
9. If the claimant has proffered a new explanation, does it account for as many phenomena as the old explanation?
10. Do the claimant's personal beliefs and biases drive the conclusions, or vice versa?

Shermer rates items on a scale of .9 (highest) to .1 (lowest), according to their level of validity within each boundary area:

"NORMAL SCIENCE"
Heliocentrism: .9
Evolution: .9
Quantum mechanics: .9
Big Bang cosmology: .9
Plate tectonics: .9
Neurophysiology of brain functions: .8
Punctuated equilibrium: .7
Sociobiology: .5
Chaos and complexity theory: .4

"NONSCIENCE" (OR PSEUDOSCIENCE)
Creationism: .1
Holocaust revisionism: .1
Remote viewing: .1
Astrology: .1
Bible code: .1
Bigfoot: .1
UFOs: .1
Freudian psychoanalytic theory: .1

"BORDERLANDS SCIENCE" (BETWEEN NORMAL AND NONSCIENCE)
Superstring theory: .7
Inflationary cosmology: .6
Theories of consciousness: .5
Grand theories of economics: .5
SETI: .5
Chiropractic: .4
Acupuncture: .3
Cryonics: .2
Omega point theory: .1

But even this apparently even-handed ratings system is biased. Interestingly, Shermer's own beliefs are given the highest ratings. But couldn't Shermer be taking theories like "Big Bang cosmology" on faith? Anyone remotely familiar with critiques of the Big Bang would surely hesitate to assign it a .9.

In addition, Shermer heaps praise on Martin Gardner, author of *Fads and Fallacies*. A skeptical classic, *Fads* is by the same Martin Gardner who in *The New Age: Notes of a Fringe Watcher* glibly treats the Big Bang-undermining career of astronomer Halton Arp in a little more than a page of text, as if a lifetime of discordant observations could be made to magically disappear.

Shermer and Gardner share skepticism's worst tendencies: preserving their own beliefs behind a carefully applied screen of critical thinking while ascribing logical fallacies to everyone else.

Shermer's book concludes with the thorough chapter "The Great Bone Hoax: Piltdown and the Self-Correcting Nature of Science." The Piltdown man affair remains an instructive episode. From the 1912 announcement of its discovery to the 1953 revelation of the hoax, this fraudulent "missing link" rode a wave of momentum for four decades before Shermer's esteemed "knowledge filters" perceived the hoax. Shermer sums up:

>...Piltdown is a painful reminder of the fact that intelligence and education is no prophylactic against fraud and flimflam. In Piltdown we saw some of the most highly decorated and respected scientists in the world taken in by someone who was at most an amateur hoaxer. It shows that humans are pattern-seeking, storytelling animals, who seek and find patterns that fit a meaningful story. Once the story is found and a story developed around that pattern, additional confirming evidence is sought, and disconfirming evidence (or clues of a hoax) are ignored. It is a testimony to the confirmation bias, one of the most powerful explanatory models of cognitive psychologists who study flaws in critical thinking, and Piltdown

shows that scientists—even world-class scientists—are not immune. It is one thing to wonder why people believe weird things, it is quite another…to understand why smart people believe weird things. One answer is that the belief engine that drives our perceptions is so powerful that…it is almost impossible to step outside one's culture to shed the belief baggage that comes with residence in a community of believers, to filter knowledge through the belief engine in order to see the evidence for what it really is—whether that be truth or hoax.

Applying this to the Kensington Runestone—which appears to have been fully authenticated, a matter we'll look at later in this book—Shermer's writing demonstrates the skeptics' inability to see through their own biases. The Kensington Runestone was mistakenly labeled a hoax for more than twice the time it took for Piltdown—an actual hoax—to be exposed.

If the self-correcting nature of science does not work as well as Shermer would like, how much undiscovered reality languishes in "the borderlands"? And how can the skeptics be trusted to guide us through such territory?

Crashing the Subjectivity Barrier

Shermer, publisher of *Skeptic* magazine, has attempted to crash what we might call "the subjectivity barrier." In the name of clarifying that gray zone where subjectivity blurs into objectivity, Shermer has submerged himself into states of consciousness where a large element of subjective interpretation is in effect. In this way, Shermer attempts to grapple with the anomalous issues surrounding these states. Shermer has thus gone farther than most skeptics—could he go farther still?

In *Why People Believe Weird Things*, Shermer recounts his experience in an altered state caused by exhaustion during a marathon cycling event: he became convinced his support crew was made up of aliens. Shermer has also submitted to hypnosis, tried remote viewing ("psychic spying"), and worn a motorcycle helmet rigged with electromagnets in order to test the hypothesis that electromagnetic phenomena induce experiences which people interpret as religious visions, alien abductions, etc.

He also tells of an experiment with hypnosis used in conjunction with cycling (to control pain and remain focused), in which he was very resistant but eventually did achieve a genuine altered state. Eventually, he tried to be hypnotized again under controlled conditions, but the power of his rational mind was simply too great, and he does not believe he was hypnotized at that time. However, working with a woman who was hypnotized, he became convinced that the hypnotists' ability to plant suggestive thoughts in a disassociated subject was a genuine scientific phenomenon with an as-yet-not-understood neurophysiological explanation. What is the neurophysiology of hypnosis? No one knows, and therefore hypnosis is labeled a "borderlands science."

Shermer's visit to the lab of brain researcher Michael Persinger, like his hypnosis, also went only halfway.

Persinger's data correlates seismic and geomagnetic activity to UFO reports. Persinger's hypothesis is that electromagnetic effects trigger microseizures in the temporal lobes, which cause fantasy-prone people to imagine they are encountering the divine or extraterrestrial. Although clearly a gonzo area, some skeptics have looked favorably upon Persinger's work. Shermer went to Persinger's lab and donned the helmet used by Persinger to induce such experiences. As with his hypnosis, he reported some alteration of consciousness but was unable to "let go."

We are well into the borderlands of science when talking about altered states. Looking at the ticklish topic of near-death experiences (NDEs), we come across territory where skeptics fear to tread.

In *Why People Believe Weird Things*, Shermer dismisses the notion that the "tunnel of light" of the near-death experiencer is the remembered birth tunnel, as suggested by psychologist Stanislav Grof. Shermer summarizes some qualities attributed to mind-altering chemicals, noting "spiral chamber and striped tunneling effects" are common to both NDEs and hallucinogenic trips. The skeptics Schick and Vaughn rated four hypotheses for NDEs in their book *How To Think About Weird Things,* and also dismiss the birth memory hypothesis as well as the possibility of the soul leaving the body. They can the "chemical reactions in the brain" hypothesis for lack of evidence and, like Shermer, settle on Susan Blackmore's contention that NDEs are caused by a dying brain's attempt to make a stable model of reality as the lights go out.

Schick and Vaughn say "the biggest problem with the hallucination hypothesis is that it does not explain why the hallucinations at death are so similar…Until we know what chemicals are involved and why they have the effects they do, the hallucination hypothesis doesn't tell us much." And in *How We Believe,* Shermer gives a nice rap about how the "Belief Engine" becomes more susceptible to non-rational thinking under the influence of psychedelics.

But it's odd that neither of these camps make mention of the five years of DEA-sanctioned research with the weird hallucinogen DMT that took place at the University of New Mexico from 1990 to 1995. Dr. Rick Strassman, author of *DMT: The Spirit Molecule,* makes a strong case that "DMT, naturally released by the pineal gland…is an integral part of the birth and death experiences, as well as the highest states of meditation and even sexual transcendence." Strassman also links alien abduction experiences to naturally released DMT.

Strassman's work provides the link between DMT and NDEs, and DMT may very well be the same chemical released by electromagnetic phenomena a la Persinger. If the skeptics are serious about unraveling these mysteries, we suggest that a dose of DMT under controlled conditions might be the way to go. DMT would render Shermer's ability to "let go" irrelevant. DMT is illegal, but abundantly present in the plant kingdom and there are thousands of formulas for the DMT-based brew ayahuasca.

No one claims to have all the answers to the DMT mystery, but it seems to be a major component of these altered states questions. If the skeptics intend to travel in the borderlands, let them go all the way.

PART 2:
Gonzo Cosmology

CHAPTER 3

Down with the Big Bang

THE BIG BANG THEORY SUCKS. IT'S BETTER THAN THE biblical story of creation, but not by much. In fact they're not really that different: "The universe began all at once, out of nothing…"

This creates certain logical headaches. Like, how did nothing become something? Science has no good answers.

Another unattractive feature of the Big Bang, like the Bible, is the idea that the universe must end. It's morbid. Don't go there. You don't even have to go there, because the data lead to a far different conclusion: the universe has always existed, and will never die.

Several great scientists have jeopardized their careers, and in some cases lost them, by pursuing the evidence for this idea. It's like you have swear an oath to the Big Bang to get anywhere as a space scientist. Most people think science is full of fresh thinking, new ideas, and open minds. It turns out the status quo rules science like everything else, and if you want to keep your job, you better not ask any questions. Just ask Halton Arp or Fred Hoyle. These two prominent space scientists' careers took significant detours, which we'll return to throughout these pages.

What the public doesn't know is that the Big Bang theory has been seriously downgraded even among its believers. It used to lay claim to the very beginning of time and space in a fiery explosion. Now all anyone is willing to say for sure is that the universe used to be a lot smaller and hotter—kind of a step back from the days of fire and brimstone.

The Big Bangers are even saying that the universe might have been here forever, oscillating in and out, and could go on forever. Too bad for all those anti-Big Bang scientists who faced years of scorn for saying the same thing. The Big Bangers have co-opted the previously heretical ideas of their opponents. Apparently, one thing the Big Bang theory seems to do well is to surround and absorb its opposition, amoeba-style.

Was the universe really smaller and hotter? There are only three indications that this may be correct, and each of them is flawed:

Redshifts. Redshifted galaxies make it look like those galaxies are receding from the Big Bang explosion—but truckloads of anomalies exist that don't fit the standard "redshift=distance" correlation. Plus, redshifts could be generated half a dozen different ways besides in an "expanding universe."

The cosmic microwave background. Hailed as a remnant of an early hot universe, there are plenty of ways to produce this microwave glow without having to invoke a Big Bang. For instance, supernovae throw off tiny whiskers of iron that are known to thermalize and smooth out microwave energy into the observed range. The Big Bang therefore becomes superfluous in this example, which uses only well-understood physics from the laboratory—obviating the need to invoke a scientifically inaccessible "beginning of time."

The light element abundances. This refers to the ratios of the so-called light elements to each other. It is said that these precise ratios could only have come from a Big Bang. Unfortunately, there are so many "free" parameters in these calculations that you can prove anything if you move around the shells fast enough.

Since each of these three "pillars of the Big Bang" has several non-Big Bang alternatives, the primacy of the Big Bang as the sole explanatory theory is rendered dubious. There is no pea under these three empty shells. We don't believe in it anymore. We believe in infinity.

It's been a tough few years for astronomy and cosmology. These sciences seem to be lost in the overgrown thickets of the Big Bang theory.

The apparently false premise of the Big Bang—that everything began all at once out of nothing—states that the entire universe getting bigger all the time. But understand that in the Big Bang model, the distances between galaxies keep increasing, but not because they are flying away from each other with momentum from a primordial blast. Instead it is allegedly because more space is being created between them.

The galaxies are not expanding into space, space is expanding between the galaxies: this is the so-called "universal expansion."

Because most light from these deep-space objects is stretched out (redshifted), it is said that space itself is stretching as the light travels through it. The stretched light from distant galaxies is therefore said to bear the signs of travel through an expanding universe.

Big Bang astronomers (who control the profession and its publishing outlets) use the aforementioned presuppositions when they tackle the problems presented by these selfsame presuppositions. Many of these problems have proven to be intractable and every approach appears to fail. These problems would not even exist without presupposing an expanding universe.

The plain fact of the matter is that all indicators of the cosmic distance scale have always been unreliable and contradictory. This in turn casts serious doubt on the cosmic expansion rate and the age of the universe, too.

Let's investigate some prominent, newsworthy failures of the Big Bang theory in this regard.

The Key Project

One of the main missions of the Hubble Space Telescope is the Key Project, which was created to shore up the notoriously shaky distance scale and thereby clarify the universal expansion rate. But trouble broke out in 2000 when the members of the Key Project had their first papers rejected by journal referees.

This is, of course, similar to the treatment afforded dissident astronomer Halton Arp and his controversial findings. The observational astronomers of the HST Key team, like Arp, made some uncomfortable discoveries and got the mad smackdown from the chalkboard theoreticians.

The Key Project studied the so-called Cepheid variable stars, which have been used since the 1920s as the first rung in the cosmic-distance ladder. The Hubble Space Telescope was able to discern (together with statistical extrapolation of its observations) that many of these stars are actually double stars, which totally screws up the distance scale, the expansion rate, the age of the universe, and a lot of tenured apple carts.

The status quo's solution to uncomfortable findings, as we know from Halton Arp's case, is to kill the messenger.

The "Twin Quasar" Paradox

Another attempt at clarifying the cosmic distance scale with the Hubble Space Telescope ran aground in 2000, this time involving the weird deep-space objects known as quasars. Analyzing Hubble images of the "Twin Quasar" in Ursa Major, a University of Arizona team managed to publish yet more evidence that the distance scale is screwy.

The "Twin Quasar" is conventionally supposed to be a single quasar with its image split into two by a galaxy that has gotten between the quasar and us. This is known as a "gravitational lens" effect. By measuring the brightness variations of the two images of the quasar, astronomers attempt to geometrically determine how far away the quasar is. This is supposed to provide the all-important universal distance scale and expansion rate.

But the accuracy of this method relies on the values assigned to the lensing galaxy's gravitational field, which had not been concretely determined in this case. And darn if the Hubble images didn't show an unusual shape for the "lens" galaxy, thus ruling out all previous calculations for its gravitational field, and consequently, all previously published values of the universal expansion based on this lensed system. More on this below.

The World According to Arp

Astronomer Halton Arp claims that redshifts themselves are not good distance indicators because light can be stretched in ways besides a universal expansion. For instance, redshifted light would arise from matter freshly ejected from a so-called white hole. Arp and his supporters argue that some galaxies blow themselves apart as they eject pairs of redshifted quasars from central white holes.

This is of course precisely what we see in the case of the "Twin Quasar": a peculiar galaxy with a pair of high-redshift quasars aligned across it. It sounds like a classic "Arp object" to us. It is therefore probably not a single distant quasar being lensed by a foreground galaxy, but rather a pair of anomalously redshifted quasars being ejected by a galaxy bursting open like a seedpod.

And if that is the case, then redshifts do not indicate distance after all and the universe is not expanding.

Bright and Iron-Rich Quasars

In 2002, the discovery of some extremely bright quasars, and of an iron-rich quasar, caused a kafuffle. Quasars are supposedly hugely distant remnants of the universe's earliest, iron-free epoch, so these observations are giving mainstream astronomers and cosmologists a real headache. The Big Bang theory requires that quasars (supposedly primitive galaxies) be far away, and therefore dim—but they are routinely far too bright. No one in the mainstream has the guts to defy their elders and state the obvious: quasars are bright because the Big Bang is wrong and quasars are not that far away. And we can expect quasars to have iron because they are not from some earlier iron-free time.

The "Age of the Universe"

The BBC ran these headlines in quick succession: Dec. 1, 1998: "Universe 15 billion years old." May 26, 1999: "Universe 12 billion years old." Feb. 19, 2000: "Oldest quasar 13 billion years old." Each headline dismisses the previous one. This has been going on for a long time. And it continues: in 2002, a study published in *Science* announced it had narrowed down the age of the universe to between 11 billion and 20 billion years old. And round and round we go.

Different methods of calculating the age of the universe create different results. The range of possible ages is large. We're not quibbling about a billion years or so. For decades, the field has remained wide open between 10 billion years old or 20 billion years old.

The age of the universe seems to depend on whom you ask—a cosmologist or an astronomer. Most of the cosmologists are running calculations and simulations that find a more youthful universe. The astronomers, on the other hand, are always guessing the age of the universe to be closer to the high end of the age range. It seems that every time a cosmologist announces a definitive age of the universe, based on elaborate calculations, an astronomer finds a quasar older than that.

This is reminiscent of the days when the astronomers insisted that the solar system could not possibly be older than a couple of billion years; meanwhile, the geologists knew that the Earth itself was a couple of billion years older than that. This matter was politely dropped when the astronomers sheepishly took it all back.

In the current conflict over the age of the universe, both sides are wrong. The age of the universe is never going to work out because the universe has no age. It is infinite in duration and has always existed. Every scientific alternative to the Big Bang theory, and there are plenty of them, agrees on this point of infinite duration.

Galaxies

Not only is age a problem, but so is the existence of galaxies. According to Eric Lerner's book *The Big Bang Never Happened,* there are structures in the universe so large that it would take far longer than 20 billion years to create them. We speak here of the monstrous walls of galaxy clusters strung about the universe like party streamers. More of them are turning up all the time. Everywhere you look, there are just huge clumps and clusters of galaxies separated by gigantic volumes of empty space. This directly opposes one of the fundamental expectations of the Big Bang theory, which is that the universe should be homogenous and well mixed like a good martini.

Astronomers have tried everything to make this anomaly vanish as a statistical quirk. They've studied the sky in wide-ranging, shallow surveys out to 200 megaparsecs, and in narrow-but-deep surveys out to 1000 megaparsecs. But no matter how you slice it, the universe comes out chunky, even at the highest redshifts representing the earliest universe. These early-forming giant structures remain unexplained.

As soon as you remove the time constraints, of course, forming these giant structures can be done in quite a leisurely fashion. They have literally all the time in the world to do it.

Outrageously, galaxies themselves are anomalies according to the Big Bang theory. All manner of theoretical wrangling and ad hoc assumptions have to take place in order to accommodate these fundamental units of cosmic structure.

Galaxies are ostensibly the result of imperfections in the early expansion of the universe which—to accommodate the existence of galaxies—is supposed to have taken place in fits and starts. This is the corner of the Big Bang known as Inflation theory.

And then, to accommodate various other anomalies associated with galaxies in the Big Bang theory, we come to the corner of the Big Bang known as Dark Matter theory. Dark matter is either hot, warm, or cold, but each option is saddled with insurmountable theoretical troubles.

Perhaps the trouble is the Big Bang itself. Its labyrinthine coils grow more unwieldy with each surprising find, and each subsequent layer of theoretical abstraction is smoothed on like balm. The theory branches off in a thousand unresolved forked paths, providing no clarification of the phenomena it purports to study. No wonder cosmologist Fred Hoyle said that the Big Bang is less like a coherent theory and more like a gardener's catalog.

The Big Bang's missteps, such as those outlined above, expose a troubled theory. It's the kind of trouble that visits theories near the end of their lives, when they are in decline. Every new result arrives completely unexpectedly and totally confounds the scientists involved. This means the theory has virtually no predictive power, and will quickly be abandoned when it becomes expedient to do so.

Why doesn't the Big Bang just dry up and blow away? The reason, as explained by T.S. Kuhn in his book *The Structure of Scientific Revolutions*, is that old theories do not disappear until there is something definite to take their place. As it stands now, the anti-Big Bang community is somewhat divided over what the successor theory should be—although they all agree it will include infinity of time and space. A hybrid theory of some kind could be just the ticket. But you can rest assured that as soon as the dissenters can agree on the best replacement, the Big Bang theory will be stuffed into the nearest trashcan. It is so tattered and torn right now that the *only* thing keeping it from the dustbin of history is the lack of a clear successor—although there are plenty of good candidates.

CHAPTER 4

Big Bang Alternatives

JUST AS THE BIG BANG IS LOOKING DULL AND WORN OUT, a number of scientists have come up with refreshing alternatives. These theories get cool receptions in the halls of academe and the musty corridors of the journals, but informed dissent is making inroads nonetheless.

The cigar-chomping space scientist Geoffrey Burbidge, a longtime Big Bang dissenter, has noted that most of those who tangle with the Big Bang establishment are older scientists with long track records and probably tenure. These are the ones who are not afraid to lose their jobs by speaking out and following their noses into forbidden research.

The notable exception here is Halton Arp, who, as recounted in his important book *Quasars, Redshifts and Controversies*, lost his job anyway. His crime? Insisting that certain anomalous observations needed to be followed up. His telescope time was terminated, ending one of the most productive careers in astronomy.

Don't think for a second that this didn't send a message to all the young grad school punks who might have been looking up to Arp. The message was clear: don't stray from the path—even Halton Arp, author of the landmark *Atlas of Peculiar Galaxies*, isn't immune from official censure.

The Anti-Big Bang Bunch

In this corner, reigning champion and dominant scientific theory of the universe: the Big Bang theory. And in the other corner, the challengers: the Quasi Steady State Cosmology, the Plasma Cosmology, the Meta Model, and Halton Arp's Continual Creation theory.

It's like Galactus vs. the Fantastic Four. Galactus (the Big Bang) is massive, overwhelmingly powerful, and has the world in his grip. The Fantastic Four, on the other hand, are freaks with weird powers who have no chance.

Like the Fantastic Four, many of the alternative cosmologists are old friends and colleagues, and provide support for each other in various ways. They are all highly decorated professional astronomers who frequently have trouble getting papers published in the major journals of their field (except the plasma physicists, who have their own field).

What unites these varied theories (besides their loathing of the Big Bang) is their acceptance and defense of Arp's heretical—yet empirical—observations of anti-Big Bang phenomena.

Before his heresy got him banished from the best telescopes in the world—Mount Wilson and Palomar observatories—Arp specialized in the so-called "peculiar galaxies," which don't fit into any of the clear-cut galaxy types like barred, spiral, elliptical, etc. What Arp noticed was that extremely weird galaxies often look precisely as if they are exploding, ejecting quasars and other weird space objects. Now, galaxies are not supposed to eject quasars; in fact this is impossible, according to the Big Bang people. They say that quasars have such high redshifts that there couldn't possibly be an association with mere galaxies, which have low redshifts.

Arp and his supporters argue that these are no "mere" galaxies, as they are often in the throes of patently violent explosive events. And you can bet there will be multiple quasars in the same area, sometimes connected by filaments of plasma, sometimes fired out in long lines. This type of observation has been given a name: "the Arp effect."

However, Arp's own ideas about what he's seen are so extreme that he is a pariah even among some of his supporters, who take what they can use from him, but can't quite commit to following him all the way down the rabbit hole—essentially, a complete rejection of the redshift-distance correlation.

The redshift-distance correlation is something the Big Bangers have on their side. It really, really looks like the higher the redshift an object has, the further away it is. But then Arp and the rest of the bunch say it really looks like these different redshift objects are physically interacting. Each side produces its own statistics to prove its own point. And there it stands, with the dissenter smacked down like Galileo.

So, what are these ragtag anti-Big Bang theories that support Arp?

The Plasma Cosmology

Plasma physicists support most of Arp's observations, and he speaks at their conventions. This is good because the mainstream of his own field (observational astronomy) has stamped out his voice at conventions and in journals. Anyway, the plasma physicists claim to have beaten the astronomers at their own game while playing on an entirely different field. By downplaying the importance of gravity and placing greater emphasis on the interaction of large-scale electro-dynamic forces, the plasma physicists interpret all phenomena through a vastly different lens. According to them, there is a perfectly rational explanation for all the phenomena of space using the parameters of plasma physics alone. Without any of the assumptions of the Big Bang such as a beginning of time, how galaxies were formed, or redshifts—you name it, and the plasma physicists have an alternative explanation for it.

Plasma cosmology is an incredibly formidable challenge to the Big Bang because it comes from outside the field of astronomy. It's science vs. science. We think it's nice to see some new faces.

We'll return to the gonzo topic of plasma physics in Chapter 11.

The Meta Model

The Meta Model is the brainchild of Tom Van Flandern, a loose cannon rolling across the decks of astronomy. Routinely excluded from conventions and journals, Van Flandern supports most of Arp's observations and sells Arp's books next to his own at his website, Metaresearch.org. He and Arp even co-authored an anti-Big Bang paper.

In a way, Van Flandern is even further out than Arp. Not merely content to dispense with the redshift-distance correlation, Van Flandern (and a small coterie of conspirators) proposes alternatives to gravity, black holes, and relativity. This is truly gonzo stuff, and worth a look. Of particular note is Van Flandern's idea that not only are space and time infinite, but *scale* must be infinite too. From the ultra-nano to the mega-macro, it just keeps going to infinity up and down the scale.

In Van Flandern's conception, the universe really is a grain of sand.

The Quasi Steady State Cosmology

The theory of the Quasi Steady State is that the universe continually unfolds in explosive "matter-creation events" from the hearts of galaxies, and that these "little bangs" drive a kind of oscillatory universal expansion. (This is of course in opposition to the Big Bang theory, in which a single matter-creation event caused the universe to expand.)

The Steady State cosmologists—Fred Hoyle, Geoffrey Burbidge, and Jayant Narlikar—are an interesting bunch. Like Van Flandern, they all support most of Halton Arp's radical claims, and they have all worked closely together in the past. The late Fred Hoyle, who was knighted some time ago for his distinguished scientific career, was one of the Big Bang's greatest detractors. Ironically, he coined the term "Big Bang"—but as a term of derision. Jayant Narlikar, author of *Violent Phenomena in the Universe*, is an ally from the formidable ranks of Indian astronomers. Narlikar has also authored papers with Arp. And Geoffrey Burbidge is in part well known for his long marriage to Margaret Burbidge, a glass-ceiling-smashing female astronomer. She is an unofficial member of the bunch, because everybody knows she's there, but she never signs the papers—the Invisible Woman of our Fantastic Four analogy. Geoffrey Burbidge has directed prestigious observatories and is himself a highly decorated astronomer. Burbidge is a long-time Arp defender, chomping on a cigar and penning reviews to Arp's books where he writes things like, "The 'Arp effect' is only the tip of the iceberg."

The Continuing Creation Theory

Of all the anti-Big Bang theories, Halton Arp's theory of Continuing Creation is perhaps closest to the Quasi Steady State Cosmology. However, Arp no longer accepts redshifts as cosmic distance indicators at all, and is therefore led to the view that the universe is not expanding whatsoever. (In the Quasi Steady State Cosmology, as in the Big Bang theory, the universe is still expanding.)

To Arp, the only means of determining where things are situated in the sky is through their apparent optical and statistical associations with each other. According to Arp, if two objects in space appear close together (in terms of "arc seconds"), and if they fall within a certain pattern, then the association is real no matter what the objects' respective redshifts.

This view completely rejects the notion that we can see very far into the universe, period. For Arp, essentially all extragalactic objects (including the Milky Way) belong to the Local Supercluster of galaxies. Anything beyond this limit is essentially unknown and as yet unseen. In this, Arp goes way beyond even what other Big Bang detractors are willing to admit. Most of them still believe, at least in some limited cases, that redshifts are capable of reliably indicating distances.

As much as Arp's observations are needed by all of alternative cosmology, his complete redshift rejection unnerves them. From Arp's book *Seeing Red: Redshifts, Cosmology, and Academic Science*:

> The best scientists I had personally worked with,
> and valued personal friends, [paid me a visit]—
> Fred Hoyle, Margaret and Geoffrey Burbidge and
> Jayant Narlikar. It was nice to have them
> there...With great zest I laid out before them
> what I had discovered...They were
> horrified!...Fred said that my embracing such an
> obviously crazy result would undermine the
> credibility of our attack on the Big Bang. He was

visibly angry…It was clear that the people I
admired the most thought I was ridiculous
(p. 157).

If Arp is right about redshifts, then he is incredibly far ahead
of his time. Eventually, the Big Bang theory will be overthrown with
the help of his prescient observations. He is a full paradigm ahead
of everybody.

The "Cyclic Universe"

"Perhaps the [recently proposed Cyclic Universe] model should be
called an old paradigm since it reinvigorates ancient cosmic
mythologies and philosophies…"

The unfortunate quote above is from a 2001 paper called
"The Endless Universe: A Brief Introduction to the Cyclic Universe"
by Paul J. Steinhardt, professor of physics at Princeton University.
In this paper, Steinhardt introduced his Cyclic Universe theory,
which he co-proposed with Neil Turok of Cambridge.

Steinhardt says that in his and Turok's new/old paradigm, "space and time exist forever. The Big Bang is not the beginning of time. Rather, it is a bridge to a pre-existing contracting era.
The universe undergoes an endless sequence of cycles in which it
contracts in a big crunch and re-emerges in an expanding big bang,
with trillions of years of evolution in between."

Steinhardt's grand description of his theory, and his comparison of it to "ancient cosmologies and philosophies," is woefully
ignorant to say the least, and disingenuous to say the most.
Steinhardt's description of his "Cyclic Universe" is in fact a dead
ringer for another theory, the Quasi Steady State Cosmology
(QSSC), which has received exactly no credit from Steinhardt.

We might imagine that Steinhardt would have acknowledged this similarity in his paper. But Steinhardt opted to say nothing. Steinhardt would have us cast our memories waaaaaaay back
to "ancient cosmologies and philosophies," and yet he would
have us totally forget about the QSSC, which is not even two
decades old.

We E-mailed Geoffrey Burbidge, one of the co-authors of the QSSC, to ask what he thought of the Cyclic Universe theory and its rather obvious similarity to his own.

He E-mailed back, saying with barely disguised distaste, "Apparently the Steinhardt-Turok theory has some similarities to the quasi-steady state theory."

Burbidge also wished to remind us of the stark difference between the two theories, and wrote: "Of course the major difference is that we [steady-staters] believe that the creation process and the generation of new galaxies takes place in the nuclei of existing galaxies and is going on all around us. This is an old idea of Ambartsumian and there is lots of observational evidence in favor of it which is discussed in our book *A Different Approach to Cosmology*."

Burbidge apparently believes that even though the Steinhardt-Turok theory of the Cyclic Universe has a cosmetic similarity to the QSSC, it remains a mere Big Bang variant, with most everything intact except a beginning of time and space.

Burbidge, the long-time Big Bang detractor, likely finds it unfortunate that Steinhardt and Turok reject a beginning of the universe in favor of an oscillation, because the Cyclic Universe thereby gains an uncomfortable similarity to the QSSC—one which both Burbidge and Steinhardt seem to wish to avoid. Does each man fear that his own theory will be undermined by its rival?

This idea took on a sinister pall when Burbidge went on to say, "Your readers may be interested in the fact that Steinhardt is a co-organizer of a meeting organized by the National Academy of Sciences in Irvine in November 2002 on cosmology and we have not been allowed to present our ideas i.e., we have been excluded from the program."

The appearance of impropriety in this case is severe. Here you have a situation where the QSSC appears to have been excluded from a conference, in part by a man promoting his own theory that bears more than a passing similarity to the excluded one. If true, that's what we would call two pounds of bullshit in a one-pound bag.

However, for his part, Steinhardt asserted in an E-mail to us that, "Yes, there is a kind of conceptual connection between this theory and ours. However, there are also many key differences...

Whether you consider ours to be a derivative of QSSC is just a matter of personal opinion. I could argue equally that it is derivative of Big Bang or Conventional Steady State [non-cyclic] or even Inflationary theory. It is, in fact, a little of each. I would say this mixed feature is what makes it something unclassifiable—something really new and different."

Yeah right. Here Steinhardt has demonstrated that particular quality of mainstream cosmologists noted by Burbidge, who said of such creatures, "Independently of anything else they are also very keen on PR."

And check out Steinhardt's criterion for getting a seat at cosmology's big table: "the criterion for serious consideration is very clear and straightforward: We begin with a standard model . . ." Ah, we get it—just agree with the standard model and you don't get excluded. Looks like all is well with cosmology!

For perspective on the politics of scientific conferences, we turned to astronomer Tom Van Flandern. Van Flandern seemed a natural choice since, like Burbidge, he is also a Big Bang detractor with scientific credentials as big as a house, and has also been excluded from a few conferences. Van Flandern had this to say: "Meeting organizers have learned to pick their attendees carefully so that they can report to the media that a 'consensus' was reached! (Such is the influence of political and advertising techniques on science these days.) And…credit rarely goes to the right people because science must maintain the illusion that all developments are progress, not setbacks. So Burbidge et al. will certainly never be credited, even if 'Cyclic Universe' eventually becomes identical to QSSC."

Then Van Flandern said something that really caught our eye: "My interest is to see if the [Big Bang] becomes more vulnerable while it is under attack from the [Cyclic Universe] folks. If it does, we may have a unique window of opportunity to get [the Big Bang] replaced with something of a completely different character from either of those models."

That's when we realized that this Cyclic Universe business represents a seismically important event in the history of Big Bang cosmology. It demonstrates that members of the Big Bang establishment are getting fed up.

We went back and re-read Steinhardt's paper. We wanted to see just what inspired him and Turok to bolt from the strict confines of Big Bang orthodoxy. The answer is the newly "discovered" dark energy.

The process of its so-called discovery has been full of participants; we're not sure whose idea it was first—might even have been a team, we don't know. Now everyone in the establishment just accepts it as fact even though it is clearly the ridiculous result of a conceptual error. This dark energy had to be invented to explain the completely surprising result that the universal expansion appears to be accelerating. This is a finding totally contrary to Big Bang prediction, which has it that a "big bang" would probably slow down and maybe even stop. So "dark energy" had to be summarily invented, and then used to paper over the gaping hole between theory and observation.

Van Flandern refers to these sorts of things as "theory patches" or "swidgets." They are used all the time in the Big Bang theory. It is this frequent modification, in the face of failed predictions, that critics have always held up as a sign of the Big Bang's imminent collapse.

In Steinhardt and Turok—who are otherwise Big Bang insiders—we see that "dark energy" may be the straw that broke the camel's back. Steinhardt's paper calls dark energy a "major surprise" that was "unanticipated and has no particular role in the [Big Bang] picture." He goes on to complain that "the general view has been that it can simply be added by fiat to the initial make-up of the universe."

With dark energy then, we finally have something too ad hoc for even conventional Big Bangers.

Inspired by Van Flandern's comment about the Big Bang's possible vulnerability at this moment, we asked Steinhardt: "Is there a sense in which the Big Bang is in trouble, such that it needs your Cyclic Universe theory to put things right again?"

After what we perceived as waffling, he eventually came right out and said, "What may be in trouble is the [Standard Model] idea that the Universe had a beginning followed by inflation [inflation being an earlier theory patch which no one seemed to mind since it smoothed over some troubling observational discrepancies]. This theory looks increasingly complicated now that we know we must add dark energy to the story. It means we now need two extra ingredients, each of which must be finely tuned. This opens the door for a new approach which is more parsimonious, which is what our [Cyclic Universe] theory is proposed to be."

In other words: yes, the Big Bang theory is in trouble. It's official now. Steinhardt's solution was to take a step closer to the QSSC by offering the Cyclic Universe theory instead. And only going partway enables Steinhardt et al. to make many more converts.

May we suggest, however, that going partway only keeps the Big Bang's problems partly solved. Reject the Big Bang completely.

CHAPTER 5

Interview with Astronomer Tom Van Flandern, Part 1

TOM VAN FLANDERN IS A SCIENTIFIC ONE-MAN BAND. He earned a doctorate in astronomy from Yale in 1969, specializing in celestial mechanics. He has held positions at the U.S. Naval Observatory, Jet Propulsion Laboratory, University of South Florida, University of Maryland, and the Army Research Lab. He is currently a published author, journal editor, online borderland science guru, and leader of the astronomically significant worldwide expedition known as "Eclipse Edge Expeditions." You can find out more about him at his website, Metaresearch.org. We conducted this wide-ranging interview with Van Flandern via E-mail in October 2001.

GONZO SCIENCE: In your opinion, what are the Big Bang theory's most obviously flawed assertions or presuppositions?

VAN FLANDERN: a) Redshift can be caused by any of nearly two dozen mechanisms. The premature assumption that the cause was expansion for reasons that would be judged invalid today was, in my opinion, a major wrong turn for the whole field. b) Microwave radiation can result simply from the intergalactic medium becoming as cool as possible, given that it is bathed in the energy of distant starlight. That happens to be about 3 degrees Kelvin, the same as observed. Jumping to the conclusion that this radiation was the predicted (at higher temperatures) relic of the Big Bang fireball was also premature. c) When galaxy rotations and clustering did not behave in a Newtonian way, invisible "dark matter" was invented and imagined to be distributed as needed to make everything work as expected. That slippery slope of inventing imaginary things has continued to its ultimate form today in the equally hypothetical and unobservable "dark energy." Yet classical physics demands that Newtonian gravity must change form at some scale, and the observed "anomalies" are just as expected by this classical model.

GONZO SCIENCE: Besides the "big names" who are on record as opposing the Big Bang theory, what can you tell us about the state of dissatisfaction in the scientific rank-and-file?

VAN FLANDERN: Many astronomers outside of cosmology don't like the Big Bang, and readily see its unscientific elements (such as needing a miracle beginning, and making the dimension of time finite). The public at large is apparently extremely skeptical. An informal poll of about 500 attendees at a debate two years ago on the merits of the Big Bang showed no support whatever for the idea that either leading Big Bang variant would still be on the table 80 years from now. But cosmologists who expect to earn a living and get published have little choice in the matter, and the same is true of science writers who depend on contacts in the field for their stories.

Several critical tests of the Big Bang are now in progress. One of those involves high-redshift-quasar proper motions—a sure sign that the redshift-distance law is wrong if they exist. The cumulative weight of such results will take a toll on the field as it becomes ever more evident that cosmologists are basing too much on unverified assumptions and adding too many untestable free parameters. But as with most paradigm shifts, they can never be complete until the last of the old breed dies out.

GONZO SCIENCE: Your Meta Model cosmology is fairly complex. Could you sum up your cosmological ideas for the layperson?

VAN FLANDERN: In my opinion, there is only one difficult concept in the model—infinity. It is notoriously difficult for finite beings to think in terms of the truly infinite. But it can be done using techniques taught to us by Gamow. For example, resolving Zeno's famous paradoxes (e.g. those proving that motion is impossible) requires infinite divisibility of matter and plays a key role in the Meta Model development.

The Meta Model differs from other cosmologies in that it is a completely deductive model whose starting point is an assumption-free void with no physical properties. When we derive properties (space, time, mass) deductively instead of adding them by assumption, we see subtle but important differences in their meanings. These differences turn out to be important for resolving paradoxes down the road.

An overview of the results of this deductive process is that the universe is infinite in time, space, and scale (not assumed—deduced); that everything is infinitely divisible and can be assembled infinitely, too; and that every element of space exists because it is occupied at some infinitesimal scale. The model also contains a complete particle model for gravity and relativity, and points the way to resolving quantum paradoxes. In particular, the universal speed limit is gone from the model, as is "heat death" (always-increasing entropy).

To me, the most satisfying part is finally providing an answer to the age-old vexing question about how the universe can be, without need of a miracle beginning.

GONZO SCIENCE: Has the Meta Model made any headway against competing cosmologies? What alternative is best positioned to take the place of the Big Bang? Could the anti-Big Bang cosmologies ever reach a patchwork accommodation among themselves?

VAN FLANDERN: A recent conference in Cesena, Italy, showed that the Big Bang dissenters are thoroughly divided about any possible replacement model. They could not even agree on which principles of physics were valid, and which were essential. One of the main reasons the Big Bang survives is because opponents are divided in this way. No replacement model has yet gathered a critical mass of supporters to mount a challenge, and the dissident astronomers are not making any serious effort to unify their support for a competitor. Their only commonality is their opposition to the Big Bang.

[Note: A second conference in October 2002 in Sutton, Ontario, showed significant progress among the dissidents, who came to agree on a core set of physical principles.]

If research funding opened up again so that professionals could explore alternatives and still get paid, this situation would change rapidly. So the problem lies with the funding authorities, who chose to support research only on mainstream models in the mistaken belief that the answers must lie there.

GONZO SCIENCE: The Meta Model is but one of the scientific ideas you advocate. It seems like you have your fingers in a lot of pies, from the Meta Model's rethinking of the universe, to your alternative history of the solar system, to your re-imagining of the intertwined history of the Earth and Mars. How do you do it? Is it difficult to keep so many balls in the air?

VAN FLANDERN: It is time-consuming to scan the journals and read articles covering current developments in so many areas. That is why most professionals specialize so much. But I have seen the benefits of remaining a "generalist" many times. Indeed, one cannot challenge major theories at their fundamentals without addressing consequences in a vast array of impacted specialties.

Recommended Reading: *A Different Approach to Cosmology: From a Static Universe Through the Big Bang Towards Reality* by Fred Hoyle, Geoffrey Burbidge, and Jayant Narlikar; *Quasars, Redshifts, and Controversies* and *Seeing Red* by Halton Arp; *Dark Matter, Missing Planets, and New Comets* by Tom Van Flandern; *The Big Bang Never Happened* by Eric Lerner; *The Vindication of the Big Bang* by Barry Parker; *Cosmology and Controversy* by Helge Kragh; *Home Is Where the Wind Blows: Chapters From a Cosmologist's Life* by Fred Hoyle; Steinhardt's paper may be found at http://feynman.princeton.edu/~steinh/cyclintro/index.html

PART 3:
Alternative Histories of the Solar System

CHAPTER 6

The Exploded Planet Hypothesis

THE NEAR SATELLITE—THE FIRST ASTEROID ORBITER— witnessed a crime.

As soon as NEAR reached orbit around asteroid 433 Eros and started snapping photos, almost every prediction Tom Van Flandern made about its structure and features was confirmed. At the same time, almost every mainstream scientist's prediction or expectation about the asteroid was confounded. Asteroid 433 Eros is solid rock, not a rubble pile.

The media reported that although the majority had been confounded, scientific business is being transacted as usual, and promised to explain these significant anomalies in no time.

Meanwhile, Van Flandern, who has essentially solved the riddle of the asteroids, is still accorded third-party status because of his refusal to recant his heresies, and is not invited to the debate.

We will now make our own prediction: In the next 20 years, the dominant paradigm regarding the asteroids will be Tom Van Flandern's. Scientists around the world will gradually be converted by the strength of the evidence, consensus will be reached, and it will be admitted: asteroids, meteorites, and comets are remnants of an exploded planet.

The current belief is that these objects are debris from the sun's primordial accretion disk, which allegedly formed each of the planets we know today. Specifically, the asteroid belt is supposed to be an unformed planet, and this belief will be abandoned by all but a few older scientists who will eventually die. But 433 Eros is not part of a "planet that never was." It is a chunk of an unrecognized former planet, a long-lost sibling in the solar family.

A major mainstream prediction about 433 Eros was about the asteroid's structure. Since the mainstream theorists believe that asteroids are the leavings from an early solar nebula, they are wedded to a vision of multiple shattering impacts between asteroids over the eons. This would result in asteroids having a kind of "rubble-pile" structure.

According to this mainstream view, the asteroids only *look* like solid mountains of rock, several miles long, but this solid appearance is merely from settled dust that obscures the loose aggregate within. By the tenets of this theory of solar nebula formation, the "rubble-pile" hypothesis has to be true. The orbital mechanics demand it.

But, thanks to the telemetry of the NEAR satellite, the opposite scenario is now an unavoidable fact: 433 Eros is a solid rock.

This throws a huge monkey wrench into the cogs of the established theory. The NEAR scientists, and the science-beat reporters who deliver scientific discoveries to the public, are suddenly engaged in the most tremendous conceptual back flips.

Observe Andrew F. Cheng, NEAR project scientist: "It's almost impossible to imagine how it could not be a rubble pile." Yet even in the face of this excruciatingly uncomfortable fact, no presuppositions have been questioned or even examined. This is typical of paradigms on the way out. Incongruous facts, like gritty irritants in the oyster shell, are made glossy with beautifully smooth coats of cognitive dissonance.

Van Flandern takes a different approach. He knows he must apply the strictest criteria to his ideas. He understands that when a prediction is made based on a theory, then the theory lives or dies according to the results of that prediction. If the prediction fails, then the theory is "falsified."

But the mainstream does not follow through on this rule. As in the case of the "rubble-pile" hypothesis, the falsified theory is not abandoned. It's quietly amended.

In contrast, when the NEAR satellite took off, Van Flandern issued the "NEAR Challenge." Posted at his website, Metaresearch.org, the NEAR Challenge is a big double-dog dare to the rest of the astronomical community. It lays out his predictions for what NEAR would find, and most of them were proven true.

Moreover, in a move displaying incredible integrity, Van Flandern claimed in the NEAR Challenge that if his predictions were found to be untrue, then his Exploded Planet Hypothesis would be falsified. He would take it all back and not attempt to wriggle out of it by applying theoretical Band-Aids.

If only the community of space scientists who oppose Van Flandern had shown that same resolve and agreed to pit their predictions against his. Because if they had, Van Flandern would have gotten a phone call from them, saying, "Tom, we're sorry. It looks like you were right. Can you come over and help us interpret these confounding results? You can have the credit you deserve."

But Tom's phone ain't ringing. The NEAR satellite witnessed a crime all right. While it took pictures of the asteroid, Tom Van Flandern was getting robbed.

The Case for an Exploded Planet

The structure of 433 Eros is just one of the multiple anomalous facts regarding the asteroids, meteorites, and comets that indicate that at least one planet has exploded in the relatively recent solar system history. In fact, many formerly unrelated solar system oddities, when seen in this light, make sense when related to this event. This gives the theory a great degree of fruitfulness and simplicity. Astronomer Tom Van Flandern has spearheaded this alternative history with his Exploded Planet Hypothesis (EPH). His solar system is a lot busier and a lot more dangerous than what you learned in school.

Identity of Asteroids, Meteorites, and Comets

Certain facts about the asteroids indicate an explosive origin. For instance, the asteroids orbit the sun at the precise threshold of their escape velocity. The laws of physics do not allow them to orbit any faster without leaving for good. This is in perfect accordance with the alternative theory that the asteroid belt is a remnant population of ejected materials from an explosive event. An exploding planet would eject some of its mass irrevocably from the solar system, but much of the remnant debris would have speeds at just beneath the threshold of escape velocity. The asteroids are the ones that didn't quite make it out. Comets, too, display these same velocities, just short of that needed to escape the sun.

But what, exactly, are these things? There has yet to be criteria for distinguishing asteroids from meteorites, and comets—way overdue for a conceptual overhaul—should be lumped here, too, as more rocky debris.

Spectrally, comets cannot be distinguished from asteroids or meteors. All the spectrographic observations of comets have showed ices to be conspicuously absent. They aren't "dirty snowballs"—they're rocky. When you observe an object's spectra you can tell—by scrutinizing the wavelengths—just what the light has reflected off of. It's the oldest trick in the book. And while long supposed to be quite different from each other, the comets, asteroids, and meteorites display the same spectra. They even display the same albedos, which are the measurements of their reflected light, and which are different for all other members of the solar family.

It is clear that they are all quantifiably made of the same stuff; indeed, they seem to be derived from the same object.

The comets are in fact plainly indistinguishable from asteroids by all observational criteria that have been applied. The asteroids are only different in that their shorter orbits around the sun have exposed them to the sun for far longer, and their tails were burned off a long time ago.

In every respect, comets are turning out to be rocky objects of the exact same composition as the asteroids and the meteorites. They are therefore all of the same origin—namely: a planet that broke up between Mars and Jupiter. Let's review the evidence.

Comets: Dirty Snowballs?

Several anomalous facts cast doubt on the common conception of comets as "dirty snowballs." Each of these facts point to the conclusion that comets are debris from an exploded planet, raining back down on the explosion site.

Even among those who do not necessarily ascribe to the Exploded Planet Hypothesis, there is plenty of room to doubt the mainstream theory of the comets. For instance, renowned cosmologist Fred Hoyle, long known as an independent, free thinker in his field, made the following assessment in *The Origin of the Universe and the Origin of Religion*:

> What a comet is not is a dirty snowball, the supposedly respectable theory contradicted by every aspect of the approach to the Earth in 1986 of Comet Halley and by events since then. No dirty snowball at a temperature of minus 200 degrees C ever exploded as Comet Halley did in March 1991. Dirty snowballs are not blacker than jet black. On March 30-31, 1986, Comet Halley ejected a million tons of fine particles which on being warmed by the Sun emitted radiation characterized by organic materials, not dirt as one understands dirt. When heated sufficiently a snowball would evaporate smoothly, whereas Comet Halley evaporated in a series of explosions that continued as the Comet receded from the Sun long after ordinary evaporation would have ceased (p. 32).

It is plain from the above that the most basic conceptual premises of cometary nature simply do not back up the observations of comets. Therefore, it's time to consider the strongest alternative theory. In this case we believe it is the Exploded Planet Hypothesis.

Long supposed (by theory alone) to be giant ice and/or snow things from the birth of the solar system, every actual comet observation via spectrometry, and whatever else, has failed to prove this most basic assertion.

Comets look for all the world like asteroids and meteorites in the cool empirical light of the spectrograph, because they display an extremely low optical reflectivity. This is in marked contrast to the high reflectivity expected of the ices in the dirty-snowball model, which is racking up failed predictions left and right.

And not just the optical but the radar reflectivity, too, implies that comets are rocky. Comets display the low reflectivities of light and radar that are totally consistent with charred and blackened rock, as if from a planetary explosion. Their color was confirmed in late 2001, when a little robot named DS1 flew right past Comet Borrelly, a pitch-black comet.

Just like the asteroid scientists, the comet scientists need to give Van Flandern a call, so he can tell them, "Comets are dark because they were charred during their birth as chunks of an exploding planet." (Van Flandern has been saying this for years, but the comet folks were *still* surprised by the blackness of the comet. How can they be surprised when his theory predicted their results? Maybe they haven't heard of him since he is routinely jerked around by their journals and conferences.)

Comets are alleged to have formed at truly vast distances from the sun, and are supposedly the most frozen, distant remnants of the solar nebula. If this mainstream theory of the comets is correct, certain chemical abundance ratios in comets (such as $N2/NH3$) should be consistent with formation in an extremely cold environment.

However, this represents another failed prediction of the solar nebula model, as the ratio $N2/NH3$ is strangely more appropriate to formation closer to the sun, within the orbits of the planets. Clearly, if comets formed closer to the temperature of the sun, this fact would also serve to cast doubt on the conception of comets as "dirty snowballs."

If indeed comets are composed largely of ices, there is no way they could survive their perilously close swings through the sun's corona, which heats them to the tune of more than 1000 degrees. Obviously a rocky composition fares far better under these conditions. The mainstream comet experts avoid this conclusion by way of elaborate calculations, which confer tremendous fireproofing abilities to their ice comets via insulating layers of vapor. The planetary breakup model does not require such severe theoretical back flips to survive, and is therefore the simpler explanation in this case.

Cometary Orbits

Comet orbits can be roughly traced backwards in time to a common region of space—right to the asteroid belt, which otherwise represents a gap in the regular spacing of planetary orbits. Hmm.

There are too many comets in the inner solar system at any given time to support the idea that comets are primordial. This is because Jupiter's massive gravity serves to clear such debris with great efficiency. Jupiter either eats comets, which it did within our lifetime with Comet Shoemaker-Levy 9, or ejects them in a gravitational slingshot on an open parabolic orbit, never to return. Given enough time, Jupiter will wipe the solar system clean.

That's why it was such a surprise when the sun-orbiting SOHO satellite began its run and quite accidentally became the most prolific comet-discoverer of all time. SOHO blocks out the disk of the sun to get a good look at the coronal discharges, and various other things that are easier to see when the disk of the sun is removed from view. As it turns out, this strategy reveals virtually uncountable numbers of previously undiscovered comets on their way through the inner solar system.

This place is swarming with comets, a fact entirely unsuspected by the conventional model, but perfectly at home within the theory that a fellow planet has exploded in the past few million years.

If comets are leftovers from the nebular disk that condensed to form the solar system, they should arrive at the sun from both hemispheres. Then why do 84 percent of comets arrive at the sun from one hemisphere of the sky?

In addition, a nebular disk origin, or an origin from interstellar space, would produce some comets arriving on hyperbolic orbits. But not one hyperbolic cometary orbit has ever been observed. Dynamically speaking, this strongly implies a more recent origin for these bodies, to the tune of only several million, not billion, years ago.

The shape of the orbits doesn't fit the conventional point of view, either. Comet orbits are highly elongated, a fact that is unexpected for the mainstream but predicted by the planetary breakup model as the result of an explosion.

The most dramatic clue to the real origin of the comets lies in how so-called "new comets" enter the inner solar system. New comets are deduced, by various means, to be on their very first approach to the sun. These comets remain dark and tailless until they reach a distance of 2.8 astronomical units from the sun, where they "flame on" and develop their giant tails.

This is because the sun's gravitational sphere of influence begins to grow over the comet, which simultaneously acts to shrink the comet's gravitational sphere of influence. Dust and other particles are therefore free to leave the comet for the first time, and form a tail in the solar wind.

In the conventional picture, the distance of 2.8 astronomical units is quite arbitrary. However, the planetary breakup model naturally incorporates this precise distance, because guess what else lies at a distance of 2.8 astronomical units from the sun? That's right—the asteroid belt.

The breakup model says that the comets and asteroids originated in the explosion of a planet in just the orbit where we now find the asteroid belt. Those fragments of the planet that headed away from the sun return for the first time as new comets.

The new comets have never been closer to the sun than 2.8 astronomical units, because that's the distance of their planetary origin from the sun. Their gravitational sphere of influence, shrinking in the face of the sun's, begins to loosen material for the first time.

What we have here is an *a priori* condition of the Exploded Planet Hypothesis nicely agreeing with the observations. The defenders of the dominant paradigm, with the typical evasiveness of the doomed, see it as a big coincidence that comets should turn on right at the asteroid belt.

A final test may come someday when cometary orbits can all be traced backwards in time. This will be a truly epic mathematical feat that will need to correct for the orbit-randomizing effects of passing stars. However, given the evidence, we feel confident that Van Flandern's prediction on this matter will be borne out: the comets' orbits will all trace three million years backwards in time to a single point within the asteroid belt, in the former location of our long-lost planetary sibling, the exploded planet.

Cosmic Ray Exposure Times of Meteorites

Meteorites provide evidence supporting the Exploded Planet Hypothesis, too, from their age to the diamonds they contain.

The cosmic ray exposure time of your average meteorite is normally only a few million years old. This is a mere blink of solar system time, which is measured in the several-billion-year range. Since cosmic rays are eternal features of the universe, one would expect that any primordial solar system materials (such as meteorites) would display a cosmic ray exposure time of billions—instead of only millions—of years.

This reasonable expectation regarding the meteorites is definitively not the case, and therefore, meteorites are the products of more recent solar system events.

A theoretical Band-Aid applied to this dating anomaly is that multiple collisions have whittled down a large, old meteor population into a population of more newly exposed inside bits, with a shorter apparent cosmic ray exposure time. But from that idea, it follows that you would see a large spread of exposure times from billions down to millions of years in meteorites.

The opposite is true. The cosmic ray dating tells us unequivocally that meteorites are only a few million years old.

Explosion Signatures

Scientists have long known that meteorites arrive at our planet "precooked," that even before hitting our atmosphere they were blackened, partially melted, and exposed to some kind of explosive fracturing shock.

Meteorites are perfect exhibitors of the so-called "explosion signatures," which are a well-known complex of singular details amassed from the study of debris from man-made satellites which have exploded in orbit. Only explosions can create them.

Meteorite Diamonds

They find diamonds in meteorites. It happens a lot. But this genuine scientific anomaly has been paved over by the theoreticians in a maneuver known as the ad hoc explanation.

Meteorite diamonds are anomalous because meteorites are supposed to represent scattered solar system leftovers, basically congealed primordial dust grains that are billions of years old. There is nothing in this mainstream theory of the origin of the meteorites that would lead one to suppose that meteorites might contain diamonds.

Every empirical fact we know about diamonds indicates that diamonds require a high-temperature, high-pressure environment in which to form, such as the interior of a planet. These conditions are definitely not the conditions of accreting cosmic dust.

Therefore, the discovery of diamonds in meteorites presents the believers of the conventional theory with a dilemma.

One choice is that the solar nebula/accretion disk origin of the meteorites is incorrect, because it has failed to anticipate meteorite diamonds. However, accepting meteorite diamonds as a falsification of the mainstream hypothesis necessitates the acceptance of a planetary origin for the meteorites.

Ad Hoc Explanations

True to form, the establishment meteorite experts have chosen to keep their unfruitful theory (with its lack of predictive power), but have added a "theory-patch," "swidget," or ad hoc hypothesis to explain the anomaly and save the theory.

After the diamonds were found, the theoreticians worked out a way to fully explain them without sacrificing any of their theory's internal logic. They surmised that the diamonds came from interstellar supernovae grains that implanted in the meteorites as they formed in the early solar nebula.

What this swidget sacrifices is simplicity. A layer of explanatory gloss has been added. The theory lives, but grows complex and unwieldy by degrees.

Now consider the Exploded Planet Hypothesis of the origin of the meteorites. If a planet were to have formed—say, between the orbits of Mars and Jupiter—we should not be surprised if such a planet had produced diamonds in its interior. Diamonds are quite naturally expected in the context of the great heat and pressure of planetary formation. If this supposed planet were to break up, a logical consequence of that breakup would be diamond-bearing meteorites. The ideas flow smoothly from one another, and no ad hoc theory is needed.

In fact, no evidence of planetary heat and pressure found in meteorites should then surprise us. We might say that such evidence is an *a priori* condition of the Exploded Planet Hypothesis— "before the fact." This is in fact what we find.

Miscellaneous Meteorite Anomalies

Other anomalous findings stump the mainstream.

For example, most meteorites show signs of chemical differentiation, or the separation of heavy and light elements. The establishment theory does back flips to explain this, whereas any geologist will confirm for you that this process is particular to planetary bodies.

In addition, meteorites display signs of exposure to unusually large magnetic field strengths, which is not what the conventional theory ever expected to have to deal with. But a planetary origin theory for the meteorites is unphased by these magnetic indications.

Another uncomfortable find for the dominant theory is that meteorites contain saltwater. Take the above, and what you've got is a planet that exploded just a few million years ago, and we're still dodging the debris.

Orbital Dynamics of the Asteroids

Like comets and meteorites, asteroids have baffled the establishment scientists who won't give in to the EPH's charms. Take the orbital dynamics of asteroids, something that Van Flandern knows a lot about. A specialist in orbital dynamics, Van Flandern comes with a fat stack of credentials. His 1969 Yale Ph.D. is in celestial mechanics, he directed the Celestial Mechanics Branch of the Nautical Almanac Office of the U.S. Naval Observatory, and he was a consultant to the Jet Propulsion Laboratory.

Van Flandern's view is that an object's origin determines how it orbits other objects and how other objects orbit it. Therefore, the orbital dynamics of an object, or of a class of objects, can tell you the object's origin.

Any celestial mechanic knows that it is extraordinarily difficult for objects in space to capture other objects with their gravity. Collisions among objects on intersecting trajectories are quite easy to arrange, assuming the trajectories in question can intersect at all when they have all of space to choose from. But it turns out that if objects in space are not going to collide, they will almost never begin orbiting each other. The chances are so slim that the physics of the situation only allow for it in terms of probabilities that are referred to as "vanishingly remote."

Celestial mechanics allows for objects to orbit each other only if they had formed together in space. For instance, if an object develops an "overspin" condition, it causes a smaller body to split off the larger one, where it will go into orbit. Or, if a catastrophic collision or near-collision splits a chunk off a parent body, then the parent and the chunk may then orbit each other. In fact, a planet and its moons develop together from the same glob.

But the asteroids are supposed to be in a different class. They're presumed to be unformed stuff, smeared throughout space in a ring around the sun. And because they're thought to be just unformed bits, sharing the same orbital space, they cannot make moons of, and begin orbiting, each other.

But surprise: they do, and are therefore the remains of an exploded planet.

Here's how it works:

Consider the case of the early astronauts. Docking spacecraft together in Earth orbit proved to be amazingly difficult. The astronauts would get into orbit directly behind the craft they were to dock to, and then they would increase their speed. They assumed that this would cause them to approach and dock, just the way we do here on Earth.

But in space, increasing your speed not only makes you go faster, but also takes you into a higher, longer orbit. The net result is a faster speed but with ever more distance to travel. Unless you first drop into a lower, shorter orbit than the object you wish to approach, you will only fall behind. Only from a lower orbit will increasing speed bring you together—if you do it just right.

The odds of this happening are truly remote. Objects in space just don't go around capturing each other with gravity.

Take the Trojan asteroids that share Jupiter's orbit. You might think that Jupiter's massive gravity would just suck them all up. But think about this: if Jupiter is in front of one of these asteroids, Jupiter's gravity starts to reel it in, which increases the asteroid's velocity. This velocity increase translates into a slightly higher, longer orbit of the asteroid around the sun. This makes the asteroid fall behind Jupiter again, because its new, longer orbit gives it more distance to travel.

When Jupiter approaches an asteroid that lies ahead of it, Jupiter's gravity tugs the asteroid back, which slows down the asteroid. This gives the asteroid a shorter, lower orbit around the sun. This keeps it ahead of Jupiter, because it now has less distance to travel around the sun.

These orbits are stable forever without ever resulting in gravitational capture by Jupiter. The asteroids "librate" back and forth in Jupiter's orbit but are never captured.

So you can see how objects with similar orbits simply cannot capture each other—the pull of gravity on an object changes its velocity, which gives it a different, uncapturable orbit.

The upshot is that if it can be shown that asteroids have moons of their own, or are otherwise orbiting each other, then the conventional asteroid experts have a big problem. According to conventional wisdom, asteroids are unformed bodies, a kind of "planet that never was" smeared between the orbits of Mars and Jupiter. Asteroids are merely rubble piles with no tendency to have moons of their own. Even if asteroids collided, they would not form little orbiting chunks out of the resulting debris. Some of the debris would be lost to space and the rest would collapse back in on itself, resulting in a "rubble-pile" structure. And since the asteroids are supposed to have formed at the origin of the solar system, there has been plenty of time for them to collide and re-collide.

Mainstream asteroid experts were quite confident in this scenario until very recently. What a problem for the dominant paradigm, then, that asteroids have turned out to be solid chunks of rock with moons of their own. An exploded planet theory of the asteroids' origin is greatly strengthened by such observations. In fact, it predicts them.

Orbital Dynamics of a Planet's Breakup

A celestial body breaking up from the inside—as it is supposed to in the Exploded Planet Hypothesis—presents a unique dynamical situation, such that pieces of debris will gravitationally capture each other on their way out of the sphere of influence of the parent body. This is how the asteroids captured their moons.

In effect, the gravitational sphere of influence of the exploding planet suddenly vanished as it blew apart. Simultaneously, the spheres of influence of the planet's pieces were all expanding as the pieces separated from what had been the parent body. The planet's pieces found themselves in orbital arrangements that were basically clouds of debris in orbit around larger chunks. We know these chunks as the asteroids, and their accompanying debris clouds have been discovered only recently.

If it can be shown that most asteroids are actually such debris clouds, as opposed to rubble-piles, then we shall know with certainty that their origin is from a planetary breakup event. The conventional alternative—that asteroids captured satellites through gravitation—is so unlikely that the response of the mainstream has been to question the observations that support a debris cloud nature for these objects.

For years there were sporadic observations by amateurs of asteroids being briefly "occulted" (or winking out) as if they were blocked by an orbiting body. These amateur observations were easy for professionals to dismiss—until their own observations began picking up anomalies.

Take Castalia. Scientists were shocked to discover that this asteroid is actually two 800-meter asteroids resting together. This is hardly a rubble pile, and since gravitational capture in this case is nigh impossible, Castalia (the very first asteroid to be imaged by astronomers) indicates an exploded planet origin.

Then these scientists started finding asteroids with actual orbiting moons; Ida and its moon Dactyl were the first. Scenarios have been crafted to explain these unwelcome findings, but none of them is dynamically viable, and all of them are ad hoc. The mainstream is starting to squirm.

What we are seeing here is a paradigm unraveling before our eyes. When anomalies crop up everywhere you look, from the very first direct observations to the most recent ones, then the anomaly becomes the norm, and the mainstream theory self-destructs like an exploded planet.

Evidence of a Blast Wave

If the asteroid belt between Mars and Jupiter really is the remains of an exploded planet—and so much evidence indicates that it is— then we should see the evidence of a blast wave throughout the solar system. Planets and other objects would appear to be singed on one side but untouched on the other.

In the inner solar system, Mercury, Mars, and the moon all display asymmetrical hemispheric cratering. These bodies, of light-to-no atmospheres where we can see it best, are each heavily cratered on one side and lightly cratered on the other. It is precisely as if a blast wave had passed through the inner solar system.

In the outer solar system, every single atmosphere-less body (namely, several moons of the larger planets, plus Pluto) is coated with some kind of dark material whose spectra reads the same as the asteroids. Van Flandern wrote, "This may be the carbonaceous residue from the blast."

Furthermore, among those same airless moons are some that rotate very slowly, and every one of those is only coated with the black on one side, while the other side is usually icy-white. This is logical, since a slowly rotating body would naturally keep one face pointed into a blast wave for a longer time.

It is striking the way these black-and-white observations hold true in every case of airless bodies in the outer solar system. Van Flandern calls it "the Black Axiom," and it shows a real effortlessness about the planetary breakup model. Van Flandern and his Exploded Planet Hypothesis simply account for huge patches of solar system crabgrass that have never been mowed.

Multiple Exploded Planets?

Our solar system provides clues about not only one exploded planet, but many exploded planets.

Just as Saturn has banded rings, the asteroid belt is also banded. There are two distinct bands of asteroids, to be exact. Type C (for carbonaceous) asteroids occupy the middle and outer belt. Type S (for silicaceous) asteroids occupy the inner belt. Assuming an exploded planet origin for asteroids, two planets are needed to explain the different compositions of these asteroid types. In addition, there are more asteroid belts beyond the orbit of Neptune, and the evidence indicates that there are at least two of these belts. Pluto, whose status as a full-on planet is in doubt, represents the largest member of this class of trans-Neptunian asteroids. Meteorites, easily shown to be exploded planet debris, display different oxygen isotope ratios. These would seem to require multiple parent bodies, too.

Interestingly, Mars shows several signs of being a former moon of the exploded Type S planet. Mars orbits adjacent to this innermost Type S band of asteroids. Mars is a small planet whose size is not out of place among the other moons of the solar system. And it has had its atmosphere completely stripped away, as if in the face of a blast wave. Mars is incredibly asymmetrical, with the crust of one hemisphere being thicker than the other by about 20 kilometers. The two layers gradually taper off as the elevation maps of Mars make clear. This is what we would expect from close proximity to a planetary explosion: the hemisphere facing the explosion absorbed the blast and got pasted over with a thick coat of debris, while the far side was largely shielded. Also congruent with this picture is the fact that Mars seems to lack really old craters, which scientists assume to have been covered over somehow. Hmm. Plus, despite its light atmosphere, Martian wind erosion is heavy because of gusts approaching the speed of sound. Yet Mars is covered with fresh, sharply defined, uneroded craters—clearly due to the geologically recent influx of impactors from its exploded parent, the Type S planet.

The geological record on Earth is somewhat helpful in support of multiple exploded planets, too. For instance, the past five mass extinctions of the past billion years correspond to distinct geological layers. In other words, the slate is wiped clean every 25 million years or so—from the outside. Debate continues to rage in the scientific community about the origins of these Armageddons, from impactors to roving planets to our sun's dark twin. And now Van Flandern is in the mix with his self-destructing solar system.

CHAPTER 7

Interview with Astronomer Tom Van Flandern, Part 2

HERE WE CONTINUE WITH OUR INTERVIEW WITH VAN Flandern, in which he talks about his alternate history of the solar system, the Exploded Planet Hypothesis.

GONZO SCIENCE: What have been some of the most thrilling moments of your career?

VAN FLANDERN: There have been many highlights. In about 1961, I became the head of the local artificial-satellite-tracking "Moonwatch" group in Cincinnati. One night we set up to try to acquire a satellite launched less than two hours earlier as it completed its first orbit, thereby verifying that it had become a satellite and allowing accurate orbit determination. I was watching in a special wide-field telescope when the satellite came through. Because the odds were against catching it by chance in this way, I was thrilled; but more so a minute later when another satellite following the same orbit came through, and then another, and another. We saw roughly a dozen that evening, telegraphing the results to Moonwatch headquarters in Cambridge, Massachusetts. It turned out that the rocket had exploded in orbit, and we were the first to realize that and get observations of some of the major fragments.

My colleague David Dunham realized the interest and value of grazing occultations of stars by the moon. If the observer were located in just the right place, the star would disappear and reappear many times as it grazed past mountains and craters at the moon's limb. I was one of the first to realize that the same mechanism would make solar eclipses more interesting and spectacular if observed from near the path edges. That concept was confirmed in 1970 when I saw the most spectacular eclipse of my life. I have since formed Eclipse Edge Expeditions to take interested members of the public to the places where they can see the most spectacular views of solar eclipses. None of the commercial eclipse tour groups have yet caught onto this idea, although some veterans from past Eclipse Edge Expedition groups have.

In 1978, a colleague and I observed a near-occultation of a double star by asteroid 18 Melpomene at the U.S. Naval Observatory in Washington, D.C. During the most critical minute, we recorded rapid scintillation in the star's light lasting about three seconds, which apparently was caused by debris in orbit around the asteroid primary. This and direct observations of occultations by asteroid satellites around this time led to the first realization that asteroids were complex dynamical systems with substantial mass in orbit around the primary nucleus. I also realized the implications of this almost from the start—that asteroids had originated in a planetary explosion event that produced debris clouds initially around all asteroids, because satellites cannot be produced by capture or by collision under any likely circumstances.

In the past 25 years, most of my exciting moments have been on the theoretical side. In 1976, I came across the strongest evidence yet that the Exploded Planet Hypothesis is a factual part of solar system history. In the 1980s, I came upon an explanation for the "origin" and nature of the universe that required no miracles. In the 1990s, I found conclusive evidence that Mars was a former moon of an exploded parent planet.

However, probably the most dramatic moment of my career came in December 1996, when I came upon what was for me compelling evidence that the Face on Mars is an artificial structure. I say "dramatic" rather than "thrilling" because, frankly, this knowledge came as a shock to me, and I went around in a daze for a while, trying to assimilate this and adjust to the new reality it implied.

GONZO SCIENCE: What was it like to start out in your field? When did you realize that your views were beginning to significantly diverge from the views of the scientific orthodoxy?

VAN FLANDERN: In the early part of my career, my work and research papers were widely read and accepted. I never had the experience of a rejected paper. However, I had my first taste of what it meant to find something incorrect for the times in 1976 when, while trying to disprove the unpopular Exploded Planet Hypothesis, I came across strong evidence confirming it. I found myself attacked, and even ambushed at a presentation before experts at an International Astronomical Union colloquium. During the next several years, I repeatedly heard the advice that persisting [in defending the EPH] would damage my reputation and career. Naively, I kept returning to the evidence, expecting my colleagues to address its merits. I now know that merit pales in comparison with the interests of experts to remain the leaders of their fields; to keep their books, papers, research grants, and teachings from becoming obsolete; to keep intact that which makes them feel special. The merits of the evidence were never a serious factor in subsequent developments that eventually did end my career as a respected mainstream astronomer. This forced me to the "fringes" by definition—areas of astronomy not accepted as credible by the experts of the field.

While sitting on the fringes, I did a lot of introspection about where I had gone wrong. But I finally concluded, as I wrote in *Dark Matter, Missing Planets and New Comets*, written during this period, that it was more intellectually honest to stay with compelling, unchallenged evidence than to succumb to peer pressure. I determined to see if my colleagues had any serious answers, or if everyone was being held in check by mutual peer pressure. When I found that the latter was the case, I responded by chucking all similar peer-pressure boundaries in all areas of the field, and finding— to my initial horror—that better models than the mainstream ones existed in many areas at the frontiers of the field. That led me to found Meta Research in 1991, with the goal of supporting just such research into ideas whose only drawback was conflict with a mainstream idea.

GONZO SCIENCE: Let's explore your Exploded Planet Hypothesis. Time after time, observation after observation, no matter if it's the discoveries of the NEAR satellite or the pictures of Comet Borrelly, everyone is surprised but you. Yet the mainstream theorists and the press won't give you the time of day. Is it fair to say that the only reason this theory is not widely accepted is that old bugaboo, the "lack of a mechanism" to explain how planets explode?

VAN FLANDERN: I am working on a paper containing three exploded planet mechanisms ["Planetary Explosion Mechanisms" was published in *Meta Research Bulletin*, vol. 11, no. 3, Sept. 15, 2002. It discussed the three mechanisms: phase changes, natural fission reactors, and gravitational heat energy]. But no, I don't think the lack of a physical mechanism until now means much. We still have no satisfactory theories for supernova explosions. We have many models, but each of them requires one of those "miracle" steps, usually at the point of getting the whole star to collapse and explode at once.

The real problem with acceptance is that the EPH model replaces more than a dozen other models of the field, each a specialty field with its own experts. In a sense, the expertise of these experts, the thing that makes them feel important, would be undermined if the EPH were right. Several have told me in frank discussions that they would leave the field and do something else for a living if the EPH were proven correct. So the main problem is the resistance of the people in the field to such fundamental and far-reaching change. It's okay to make incremental progress, but paradigm shifts affect too many careers and livelihoods.

GONZO SCIENCE: Some of the ideas you champion, notably the Exploded Planet Hypothesis, seem like they could almost be accepted by the scientific community tomorrow. Conversely, some other theories that you stand behind (like the artificiality of the Face on Mars) have been so ridiculed and maligned that they may never get a fair hearing. To what extent do you worry that the EPH may be denied acceptance because of your association with yet more controversial ideas?

VAN FLANDERN: When I helped start Meta Research in 1991, we realized that we would be specializing in things that do not fit the mainstream models throughout the field. Our [mainstream] colleagues tend to think highly of our work in areas distant from their own specialties, and lowly of our work that impacts their own specialties. In other words, the same sort of dilemma (some might accept this if it weren't for that) is a recurring theme throughout everything we do. So we simply apply the same set of standards rigorously across the board, and let nature direct us rather than human politics.

GONZO SCIENCE: Would you like to comment on the degree to which your seemingly disparate group of theories actually inform and support each other?

VAN FLANDERN: They have in common a deductive methodology and strict adherence to the anti-bias protocol of scientific method. Everyone knows that scientific method demands that theories be testable; but most have forgotten that biases will more often than not lead test results to confirm a pre-determined outcome, unless controls against bias are in place.

In some cases, such as the EPH, a single model replaces dozens of others. In that sense, all our solar system work is interactive and mutually supportive, as is our cosmological work. The common link between the two is gravity, and a few tests of gravitational models are in common between the Meta Model and the EPH.

GONZO SCIENCE: To what extent is scientific progress hamstrung by its own politics? What are some examples that you have seen of the politics of science gone awry?

VAN FLANDERN: Research on complementary or replacement models can no longer compete for the limited funds available, because proposals for such research cannot normally succeed in getting high ratings from a majority of reviewers who start off hostile. The result is that many bright people have been turned away because they cannot swallow some existing mainstream models; and those who do have jobs and funding in the field are selected as personalities willing to conform and submit their personal best judgments to the will of the majority. That is no way to advance science.

Gonzo Science: How would you advise young dissident scientists just beginning their careers? Should they charge into the ring and risk being unemployable, or should they compromise their ideals by keeping quiet and reform the system from within?

Van Flandern: To have an impact, one must have resources to keep researching, teaching, and/or writing. Supporting unpopular positions too early in one's career will indeed make one unemployable in the field. I see no choice for most scientists but to "play ball" while working toward a tenured position or other secure funding source, and building a reputation as a solid scientist while doing that. If one can avoid being seduced by the job security and positive feedback for conformity from colleagues, one can start to make a real difference for science sometime during mid-career.

Gonzo Science: Do you have any funny anecdotes from your career as a scientific gadfly?

Van Flandern: In so far as I know, I think I have the unique distinction of being the only member of the American Astronomical Society's Division of Planetary Sciences for whom the suggestion was made that all submitted papers by me be automatically rejected.

CHAPTER 8

Dr. Velikovsky Strikes

Dr. Velikovsky Strikes Out

During the 1950s, Dr. Immanuel Velikovsky created what was arguably the greatest kafuffle in the history of science.

With a certain degree of audacity, Velikovsky wrote an outrageous book, *Worlds in Collision*, which ignited a firestorm of controversy.

Some very big-name scientists (including well-known blowhard Harlow Shapley) organized a boycott of Velikovsky's publisher, Macmillan, which not only handed this bestseller to another publisher like a hot potato, but—to appease the scientific rank and file—they also fired the editor who had acquired the book.

The controversy over Velikovsky's ideas never quite died out, though, and so in 1974 a symposium was held in which Velikovsky was set up like a paper doll against four critics, who proceeded to stamp him out like a smoldering ember.

Carl Sagan was a prominent name among these critics, and Sagan offered an expanded version of his critique in his book *Broca's Brain*. The media reported a great victory for the scientific establishment and Velikovsky was effectively shuffled out of the public eye.

There are still journals and websites and books published by his aging disciples but no one takes them too seriously and Velikovsky is widely hated. His very name is synonymous with scientific quackery and hucksterism. It's a lost cause. He's dead now.

Why did they hate him so much? Well, he was ornery. His ego was like a steel trap. He wasn't about to take back one word of anything he ever said, which doesn't endear one to the self-correcting methods of science. Plus, a lot of his source material came from myths and legends, which he tended to interpret as literal records of celestial events.

Here's an example of Velikovsky's wack style. His reading indicated to him that in the ancient world there was confusion about the identities of the morning star and of the evening star. Both "stars" are of course the planet Venus. The Greeks called the evening star Athena and the morning star Aphrodite, after the love goddess who came to be known as Venus. To Velikovsky, this meant that Venus and Athena represented the same entity. So far, so good.

Then he turned his attention to the myth of Athena's birth from Jupiter's head, which is where his theory gets weird. According to Velikovsky's view—that myths are accurate records of celestial events—then the myth of Athena's birth must have been an ancient astronomical observation of the planet Venus erupting from the planet Jupiter, whereupon it proceeded to tear about the solar system as a giant comet before settling into its present orbit.

What Velikovsky and his disciples failed to admit is that anything can be proved using mythology and legends. It's an unscientific methodology. It's even too gonzo for us.

Dr. Velikovsky Strikes Again

Apart from myths and legends, Velikovsky's far-flung interests included topics like ancient history and astronomy. While reading old texts, he noticed a peculiar, long-ignored anomaly in the world's oldest historical and astronomical records: without fail, the ancients referred to the planet Venus as a comet. Velikovsky paired this observation with his myth-based theory that Venus had been ejected out of the planet Jupiter, a theory which is almost certainly wrong.

Despite using these shaky sources, Velikovsky absolutely flummoxed his critics with several predictions that, one by one, came true.

It seems that while Velikovsky was wrong about the origin of Venus, he had stumbled upon some legitimate evidence that Venus is the most gonzo planet in the solar system, thus utterly confounding the orthodoxy.

Back in the 1950s, when Velikovsky was writing, Venus was assumed to be a pleasant, placid little orb a lot like Earth. Maybe a little cloudy, but peaceful. It was therefore assumed that Venus's temperature was a mild-mannered room temperature.

Velikovsky enraged establishment theorists when he asserted that Venus wasn't meek and mild—it was a fiery young planet. Once the mainstream scientists quit snickering they found that Venus was geologically pristine, with almost no craters—certainly no old ones. It also turns out that the temperature on Venus is hot enough to melt lead. In fact, since then, Venus has been described as "a giant volcano." All of this is what you might expect from a brand-new planet.

Velikovsky was also right about Venus's rotation and orbit. His reading indicated to him that Venus had made a close pass by Earth before it settled into its present orbit. He therefore predicted that Venus's rotation would be disturbed.

When it was investigated, Venus was found to have a messed-up rotation. That's right—Venus rotates backwards, which means it is upside-down.

And, what's more, its orbital period is exactly 5/8 that of Earth, which means that every four years, at its closest approach, Venus turns the same face to us at the same moment. This is called a resonance lock, and is enormously supportive of Velikovsky's theory. It is almost certain that Venus and the Earth used to be closer together. Velikovsky didn't predict the resonance lock *per se*, but can you imagine the look on his face when it was announced? Everywhere he looked, the evidence supported him.

Velikovsky's heresy didn't stop there. Not only did he loudly claim that Venus was a planetary newcomer, but he insisted that gravity alone was insufficient to explain the dynamics of the solar system and the universe. Electromagnetism must also play a fundamental role, he maintained. In the 1950s this was sheer poppycock, but of course now we know he was right.

For instance, the astronomical theorists of the day assumed that Jupiter must have been a cold, silent world. Velikovsky predicted that Jupiter would be found to emit radio noise as a consequence of being a vibrant, electromagnetic body.

When this one came true, renowned sci-fi prophet Arthur C. Clarke sent him a telegram that read: "Dr. Velikovsky strikes again!"

A great little anecdote must be related here. It seems Velikovsky had Einstein's ear, and the two men would chat occasionally. This isn't so strange, as it's often noted that Einstein was very generous to the dozens of crackpots who sought him out to validate their pet theories. The skeptics lump Velikovsky into this camp, of course. Einstein was quite skeptical of Velikovsky's theories, too, until the news broke about Jupiter's radio noise.

This latest success so impressed the old Jedi that it caused him to reread Velikovsky's *Worlds in Collision*. Eight days later, Einstein died with the book open on his desk.

What no one can figure out is how Velikovsky derived so many correct predictions from his crackbrained theories. The obvious answer, of course, is that at least some of his theories aren't so crackbrained after all.

There are actually two observations in the astronomical literature—one from the 1500s and one from 1686—of Venus displaying a weird comet-like appearance. It's called the Maedler Phenomenon, and Velikovsky doesn't seem to have been aware of it. What is its origin? What process does it signify? And how did its rediscovery by Velikovsky in the records of ancient man lead to so many correct predictions?

No establishment scientist dares to look into it, for Velikovsky's name is still the kiss of death.

CHAPTER 9

Interview with Velikovskian Researcher Charles Ginenthal, Part 1

CHARLES GINENTHAL, AUTHOR OF *CARL SAGAN AND Immanuel Velikovsky,* co-author of *Stephen J. Gould and Immanuel Velikovsky,* and editor-in-chief of the journal *The Velikovskian,* spoke with us during the summer of 2002 from his office in Forest Hills, New York.

GONZO SCIENCE: Could you give us an overview of Velikovskian theory?

GINENTHAL: What Velikovsky said is that about 10 or 12,000 years ago, probably 10,000...Jupiter became unstable. A great deal of material existed in the solar system because of a previous nova-like event, and as this material fell into Jupiter—this is the theory of Raymond Lyttleton, this is what [Velikovsky] took to explain this—that as material fell into Jupiter, it fell on an angle, and it created greater rotation of Jupiter. And if you get Jupiter to rotate fast enough—it's rotating quite quickly now in about nine or ten hours. It rotates extraordinarily fast, thus it would become unstable. And to offset this instability, it would throw off pieces of itself. We are suggesting one of the pieces came from the core, which many astronomers say is made up of Earth-like materials. Therefore [this piece of] core would leave Jupiter and be ejected into space. It would come close to the Earth in time—over a long period of time, eventually it would come close to the Earth.

But since Mars was an inner planet at the time, according-ly, then it would have also come close to Mars. And Velikovsky describes Venus coming close to Mars as a great war between Venus and Mars, and Mars then being ejected from its orbit while Venus was settling into its own orbit. And therefore Mars was not a comet, it was an inner planet that was ejected from its own inner orbit.

GONZO SCIENCE: Unlike Venus, which [in this scheme] began as a giant comet.

GINENTHAL: Began actually as a giant planet that was hotter than the sun, from its birth, that is, from the inner core of Jupiter, which is hot. This, to me at least, makes reasonably decent sense.

GONZO SCIENCE: And where could one find support for these ideas?

GINENTHAL: Well, not only in the myths, but in what we find on Venus…That is, the gasses are young. It has two gasses—which totally contradicts the idea that Venus is old—argon-36 and argon-40. Argon-36 is an unstable form of argon.

Carl Sagan himself, who was a critic of Velikovsky, said that Jupiter is a remnant of ancient times and its gasses would reflect what was going on in ancient times. Its gasses can't escape. And argon-36 on Venus was stated to be, by the astronomers themselves, that of a newborn planet.

Now, the problem they had is that they found there was a great dearth of argon-40, which is made up of the breakdown of potassium-40. If Venus is as old as the Earth, then potassium-40 should be equivalent to what we have on Earth, or more so. What they found is that it's much, much, much, much, much, much, much less. Therefore, if you extrapolate from that, you have a young planet.

To get around this, astronomers have been playing a Catch-22 game. That is, David Morrison, a great critic of Velikovsky, and student of Sagan, says, "Look. The surface of Venus, when it was very young, literally froze up so that no more argon-40 could get into the atmosphere." And that's fine—

GONZO SCIENCE: That's kinda ad hoc, I guess.

GINENTHAL: Yes, yes, but okay—the problem is you have argon-36. That is, if argon-36 was stopped from being emitted, since it comes from the interior of a planet as well, then argon-36, which breaks down, would have three or four billion years to break down. There should be practically none of it left! The problem is that there's an enormous amount of it compared to the Earth. Not that there are enormous amounts of it, but by comparison to the Earth there's just an enormous difference…

GONZO SCIENCE: We'd always thought of Van Flandern's Exploded Planet Hypothesis as a sort of post-Velikovskian, catastrophist-style revision of solar system history.

GINENTHAL: Yes, 3.2 million years ago.

GONZO SCIENCE: Do you feel Van Flandern's theory has any merit, considering that he cites much of Velikovsky's evidence in support of it?

GINENTHAL: I have looked at his evidence. One of the things he cites as evidence for his theory are the craters on Mars from 3.2 million-years ago. But in his book that deals with the craters on Mars, [*Dark Matter, Missing Planets, and New Comets*], he also talks about something that is dear to my heart: the erosion factor on Mars. And he has the erosion occurring at a rate of about 2,000 feet or so over the 3.2-million year period after his exploding planet. And the question is: How could you have removed 2,000 feet of surface material and have the planet still looking as it does?

    The other point where I disagree with Van Flandern is about the waters that poured from the exploding planet that hit Mars and created these great big "outflow channels"—these are like the scablands of Washington. There are about a dozen of them on Mars. But these occur along a great circle arc, most of them, and they all flow back over the escarpment, between the southern hemisphere, the continent, down to the lowlands of the [northern] valley. And my question would be: Would an exploding planet create floods of the size he's describing? Yes. But would they occur along a great circle arc? That I'm highly skeptical of.

And as for the [Martian] rivers, these rivers have dendritic systems, that is, these are not the great outflow channels. They have several tributaries attached to them, and these can be seen from spacecraft, and these systems have meanders to them, and these appear to be river systems. Now, how could river systems be created by masses of water falling from the sky and flowing along the terrain? He has a strong point—and I don't know if he's made it—that the river systems don't tend to show the sharp divides that we have here on the Earth, [where] there are very sharp divides between various river systems. But if a planet is quite old and has these river systems for billions of years, then this would be an indication that Mars had water on its surface, and as I pointed out in my article in *The Velikovskian* centennial issue, the bottom of the [Martian] lowlands, which would be the ocean, has seven or eight, maybe nine minerals which are only found in ocean bottoms, as if Mars had once had an ocean. So these are the differences I have with Tom.

GONZO SCIENCE: So you're saying the water is endemic to Mars, and not from an outside incident?

GINENTHAL: Yes. If Mars had tilted, the oceans would have slopped over the lands and then flowed back to the oceans.

GONZO SCIENCE: We E-mailed Tom before this interview with you, and we asked him if he was familiar with Velikovskian material, and if he could think of any good questions to ask. He wrote back saying, "I have never understood why Velikovsky associated this comet with Venus. Yeomans back-calculated Comet Halley to the 15th century B.C., but couldn't take it any farther back because it had such a close approach to the Earth that the computer could not tell which side of Earth it passed on, and therefore what its orbit was before that event. But such a close passage of such a spectacular comet would certainly have been a major event in any recorded history of the times, and that just happens to be the origin time for Velikovsky's 'Comet Venus' too." How do you respond to that?

GINENTHAL: Well, I would say that Halley's Comet may have been a comet in orbit around Venus. And after that period, it was cast into its larger orbit by its interaction with the Earth.

We can invent dynamical theories to explain a great many things. The question is, how strong is the evidence for Venus having been a comet? Velikovsky's evidence is based on, of course, mythological evidence; I have been looking at the scientific evidence related to Venus as an extremely hot body. Velikovsky's description of Venus would be as a newborn planet that would be extraordinarily hot, highly volcanic, and with gasses that are extraordinarily young. And all of these things are found on Venus. Those who have read my book *Carl Sagan and Immanuel Velikovsky* will see that the gasses are extraordinarily young, the surface is extraordinarily young, [and] all the measurements have showed that it is hotter than can be explained by the greenhouse effect—although all the measurements were fixed up...It is the totality of these various pieces of evidence that hold together that are convincing, for me at least, and for those of us involved in the Velikovskian movement.

GONZO SCIENCE: We've always thought that the strongest evidence in support of Velikovsky's theory are the weird facts surrounding the planet Venus, in particular its reported cometary appearance in the past. At the very least, Velikovsky should be given credit for uncovering this strange Venusian anomaly, even if it's only an optical effect. Do you think anything like the so-called "Venus Dog" effect, which is sort of a rare atmospheric optical illusion, or the mysterious "Maedler Phenomenon," could be responsible?

GINENTHAL: No. What I would suggest is that if Venus—and I'm using the word "if" because this is a theoretical concept—if Venus left Jupiter's core, which is nine times hotter than the surface of the sun, according to Isaac Asimov, it would have been a ball of ionized gasses. It would have extended far from what we would call the present surface of Venus.

That is one of the arguments against Velikovsky—that Venus, under its present condition, could not have had a tail. Gravity would certainly not allow it. But, if Venus were a ball of ionized gasses, and these expanded outward—that is, if you double the size of a planet with the same mass by heating it, the force of gravity at the surface falls off four times. If Venus was a hot ball of gasses, all ionized—and the ancient myths talk of Venus rivaling the sun in brightness—then if it was that hot, my estimate, and it is only an estimate, is that Venus was about four or five or perhaps six times its present diameter and was brilliantly hot, and that its surface gravity was 25 to 36 times weaker. On a surface like that, ionized gasses would have escaped. There would have been no problem with that, particularly if these gasses are extraordinarily hot.

George Robert Talbot, who makes his living as a thermodynamicist, [created a model where he] made Venus 3,500 years ago have a surface temperature between 2,000 and 1,500 degrees—I'm not sure if it was Centigrade or Fahrenheit—and he presented this material in the journal *Kronos*. What he did is what any good thermodynamicist would do: he put in all the calculations for a body roiling, bubbling with an enormous amount of heat into a very thick, hot atmosphere, taking all this into account, and tried to find out what its temperature would be. He fed in all this information [into a computer]; he didn't know the answer he would get. The answer he got is the present temperature of Venus. It's extraordinarily accurate.

Now the problem is that someone will say, "Well, he couldn't know all of these aspects of Venus." That's true. And his estimate of the temperature could be off. But the problem is that in order to prove that Velikovsky was wrong in terms of Talbot's work, you'd have to prove that Talbot was 100 percent wrong about everything. His analysis is just too neat to be dismissed out of hand. . . .

Archie Roy, the great English astronomer whose field is exact gravitational physics in the solar system, and who was very kind to Velikovsky…said he read Velikovsky's book and was very much impressed with it, but he felt that Velikovsky had gone down gloriously, because Mars, he said, could not [have been] an inner planet. And that is the problem that I raised in Italy a couple of years ago…

I raised the question of the continuously habitable zone. This is one of Victor Clube's arguments with Velikovsky. That is, the notion of the continuously habitable zone says that if you move the Earth too far away from the sun, by only a few million miles or less, the Earth will go into an ice age—and a terrible ice age. If you move it slightly closer to the sun, you'll get a runaway greenhouse effect.

The problem that I pointed out when I described all the evidence for water—that is, rivers, rainfall, and oceans on Mars, which I pointed out is in my book on the Velikovsky centennial— is that Mars [is currently] outside of the continuously habitable zone. And if it was always outside, [as] according to Roy, then how could it have had water on its surface for fairly long periods of time?

And that's what we found—that is, one of the members of the U.S. Geological Survey who passed away several years ago said that he found that a river on Mars flowed, that there was a volcanic flow that blocked a river, the river flowed over it and cut through that flow. And then there was another volcanic flow which blocked a river, and the river again flowed through and cut through that material. That means abraded it. He said that this didn't happen billions of years ago, but hundreds of millions of years ago, and maybe even closer to the present.

And so the argument is that this simply does not explain how Mars has water flowing on its surface in the form of rivers for billions of years at its present distance from the sun....

[David] Pieri, who was a student of Carl Sagan's, said it was probably water-sapping, that is, in the deserts of the southwestern United States, in the spring, the [water] from the mountains doesn't always go into rivers. Sometimes it goes underground. And it comes out in near-desert areas . . . in a rush, and it forms a river valley, and as it keeps cutting into the ground where it's coming out, it creates what looks like river systems. This was Pieri's theory.

And the problem with it is that Peter Cattermole said that this theory would be great except that in order to have this happening you need rainfall to constantly keep the water table high. So it again suggests, if that theory is correct, that Mars was not in the continuously habitable zone, and was closer to the sun. It could not have been in its present orbit and still have had rainfall for billions of years.

GONZO SCIENCE: How do you identify yourself?

GINENTHAL: I try to stay as close to Velikovsky as I possibly can.... But this is still a very long row to hoe. And basically, until the astronomy is settled, we will be considered outsiders and our work of no value.

Nevertheless, we are making our inroads, and much of what was thought about solar system development is no longer valid with what has been found with the new solar systems. The new solar systems found do not appear to be like ours. The larger gas giants are very close to their star, or they're in highly elliptical orbits.

Now, none of this makes any sense. They've tried to invent one thing after another, and one young astronomer said, "Look, we'll try anything." They'll try anything to hold on to the old theory. That is, what they've discovered doesn't match their theory. And nearly all of what they've discovered doesn't match it.

For example, they found one planet, one of these gas giant planets, with an orbit with an eccentricity of 0.67, which would be that of a cometary orbit. And it's always assumed that if you find a body going around a star in a cometary orbit, then it was captured. How do they explain it? They don't. They just ignore it. It would seem that planets are not necessarily born inside the solar system from clouds of dust. It might be that planets are captured. And that's what my theory suggests.

GONZO SCIENCE: And then their orbits are circularized somehow.

GINENTHAL: Yes, and then their orbits are circularized. Now, I'm going out on a limb, I'm not sure, since I haven't looked at the data for it, but the systems that they have with two or more giant planets tend to have planets which are in circular orbits, as if the magnetic fields of the planets are tending to stabilize them. Those with only one [giant] planet tend to be elliptical. . . . But this is conjectural, and there is no validity to it until you put up a magnetic ball [in space as an experiment] and see if what I'm discussing or suggesting is possible.

GONZO SCIENCE: And then of course once it happens and proves you right, they'll talk about how they predicted it all along.

GINENTHAL: They'll write me out of it. I've discovered many times people have made a discovery, and then after they make the discovery, somebody else gets the credit for it. Emilio Speticotto, the Italian mathematician, once joked with me and said, "There's a law called Schmeder's law. Schmeder's law says whoever invents a theory never gets credit for it." And then he told me, "Schmeder's law wasn't invented by Schmeder."

GONZO SCIENCE: What's it like to be a caretaker of Velikovsky's legacy?

GINENTHAL: It's not very pleasant at all. Tom Van Flandern takes a lot of heat for his exploding planet concepts. Victor Clube—who's such a gentleman, he dresses the way I used to dress many years ago, with the shirt, the tie, the jacket, the vest, the old Oxfordian scholar [look]—he can't get people to listen to him. I've spoken with people who agree with him, with a dendrochronologist who I meet at meetings, he puts his arms around me, gives me a big hug, and says, "You poor lost sheep!" But I see him walking around, he has his hands in his pockets, and he has this look like, "When are we going to get enough people to start to look at Clube's work?"

It's very hard being an outsider. The story that Albert Einstein was never bothered by all the abuse he took when his theory first came out is not true. He was deeply bothered by it. That's why I think he was so friendly to Velikovsky. Because he saw this in Velikovsky, and heard the arguments, the kind of vitriol that he had encountered.

They knew each other in Europe, and were friendly with each other in the United States. Velikovsky was trying to get his theory tested in some way, and Einstein said to him, "Look, if you want to make an important contribution, make a prediction that no one would make."

And Velikovsky said, "Jupiter has a large magnetic field." And Einstein asked him why. And Velikovsky said, "It has planets going around it, which are its moons, it has a large system going around it, and it should have a large magnetic field." But Velikovsky felt awkward about asking Einstein to use his influence, you know, he felt very awkward about that, it would be imposing upon a friendship. . . .

About a year or so later, they found Jupiter had a large magnetic field, which no one had anticipated, no one at all. In fact, at that time, the 1950s, they thought that Jupiter was cold inside, that it was ice layers going down, down, down. And they said something this cold wouldn't produce a magnetic field. Now Velikovsky said it produced it like the sun. He didn't say it was produced from its heat or from anything else. He said it was produced like that of the sun.

When he discovered this he brought the news to Einstein, and Einstein read the material by the people who found it, I think Burke and Penzas. He stood up, and he turned red. And he said, "What experiment would you like to have me do for you?" That is, he was ready now to go out on a limb.

Basically, what Velikovsky brought to Einstein was something very unpleasant. Einstein had lost his following to a large extent of younger scientists who had all moved over to quantum theory, which seemed to explain the universe and the world of matter much better than Einstein could. And quantum theory is based on electromagnetism. So when Velikovsky came back into his life around 1952-53—Velikovsky moved to Princeton and they met each other and renewed their friendship—here was Velikovsky saying to Einstein, "Look, electromagnetism plays a role in celestial motion." If that's so, it undermined Einstein's fundamental theory. And so here was a guy coming back who was a heretic, like Einstein was at the beginning, and bringing another theory of electromagnetism into his world.

Now a lot of people think that Einstein was thoroughly satisfied with his scientific predictions—although he admitted his personal life was a total mess. A letter was sent to Einstein by a Professor Solvine saying, "You know, you must feel gratified after so many years of looking at your work and having all of this behind you."

And Einstein said, "Where I stand, it doesn't look that good.

I'm not sure that my work will stand up in the long, long run, and I'm not even sure I was working in the right direction"....

It would seem to me that the person Einstein didn't like was electromagnetism. It was Niels Bohr, and Max Planck, and those people, and that Velikovsky had stirred up electromagnetism again. That's why I think Einstein was jolted when he discovered that Jupiter had an electromagnetic field. It was a real kick in the unconscious.

GONZO SCIENCE: And it undermines relativity because relativity relies on gravity, solely?

GINENTHAL: Yes. If electromagnetism plays a significant role, which is what Hannes Alfven is saying, then of course all the other establishment physicists are going to be up in arms...That is, what you're doing is you're questioning the holy of holies: "Gravity is the only real force in this celestial universe." But in order to solve the problems in the universe now, they're inventing something called "dark matter"—invisible matter to make—

GONZO SCIENCE: To make all their anomalies go away.

GINENTHAL: —and they just invented something else.

GONZO SCIENCE: "Quintessence"? "Dark energy"?

GINENTHAL: Yes, what is it, the energy to make the lambda theory, which Einstein called the cosmological constant, to make the universe expand. They're inventing one thing after another because the universe is not behaving the way they seem to believe it should behave.

And recently, they just received another kick, and everybody of course is looking into it. That is, they are looking at galaxies that are 12 billion light years away, and it seems that interpreting the light coming from them, the fine number constant—which is 137 in quantum theory—isn't holding up. Which is another kick in the head. I would interpret it—if it turns out; they haven't had enough people look at it to test whether it's so or not—but if it turns out to be so, it would imply that light can be tired, and that you're not losing the fine number constant: light is weakening as it travels through space.

GONZO SCIENCE: Then that opens the door to Halton Arp and the rest of them.

GINENTHAL: Yes. It opens the door to a new interpretation of the universe. The cosmologists are saying—and I see it in their books—they're saying, "We're open to new ideas, we're open to new ideas." But if you give them a new idea, you're in trouble. . . .

So you know, you asked the question before, "What's it like to carry the mantle of Velikovsky?" What's it like to be called a crackpot 25 times a month by people who won't even give you the decency of hearing you out, and looking at the evidence? The attitude is, "It can't be right, and therefore why am I wasting my time with you?"

In fact I spoke with a newspaper reporter. He had written a biography of Carl Sagan. I called him up because he attacked my work very vitriolically. And I told him, "Look, we're not all religious fanatics here in the Velikovsky camp."

Although we do have our religious fanatics—they want everything in the Bible to be just so. I mean, it's terrible. How can you carry on when Velikovsky himself said the Bible is written by men—and although he agreed with much of it, he didn't say every last iota of the Bible is true.

GONZO SCIENCE: A common misunderstanding about Velikovsky.

GINENTHAL: And to this fellow I said, "We're not all religious fanatics," and he just—you want to see what it's like to carry the mantle of Velikovsky—he said, "I'll tell you what, Mr. Ginenthal, I think Velikovsky's a nut!" What are you going to say to a guy who says that? You know, once he says that, how can you expect to have a balanced report on Velikovsky?

The reason I think that vitriol still exists with Velikovsky [from the scientific community] is essentially that there are journals. There's my journal [*The Velikovskian*], there's another journal *AEON*, there's one in England, and there are Velikovskians working all over the world doing research and writing. Velikovsky hasn't gone away. They're attempting to act as though we don't exist, and in that way, hopefully none of this will ever come out.

GONZO SCIENCE: So you would say Velikovsky is not the last of the old catastrophists, but the first of the new catastrophists.

GINENTHAL: I don't know what he is. I admire him immensely. I admire his courage. You know, none of my ideas are really my own, I come strictly out of Velikovsky. So much of his work is so original and so thought provoking. I'm constantly shocked when I do research, and find out what the research says and how it fits well with his theory. But Velikovskian theory will not be really acceptable until the astronomy—that is, the orbital work—is finally worked out.

Although—how should I put it—50 years ago they said the solar system was totally stable, and nothing held together better than the solar system. That was the work of Laplace, he said, "Planets don't change their orbits, they are held in place."

In the last 20 or 30 years, with chaos theory, a Russian by the name of Kolmogorov, with Moser, and Arnold, took the mathematics of chaos theory and began applying it to the solar system and solar system stability, and the idea that planets don't change their orbits. And they found out that there was no stability.

This was picked up later by Jack Wisdom at MIT, and he said that the solar system is not stable—but not on the short term that Velikovsky claimed. What I found wonderful about Wisdom's work, which is so gratifying to a Velikovskian out here in the trenches, was that he said that the planets with the least stable orbits are Mars and Venus—the ones Velikovsky discussed in *Worlds in Collision*. It was like manna, like water in the desert.

And retrocalculation—one of the things they [say in their arguments is that] retrocalculations prove that Venus was always in its orbit, Mars was always in its present orbit, and so on and so forth. And yet several years ago a group of experts took a look at a particular—I think it was a comet or an asteroid—and they retrocalculated it, and then calculated its orbit in the future. And what they found out was that this thing about 30 million years ago popped out of the sun, and in about 10 or 12 million more years it's going to go back into the sun, and of course this doesn't make a bit of sense—you don't get an icy ball of material or a block of stone thrown out of the sun 30 million years ago. So what they would argue in other words is that a lot of the retrocalculation doesn't work.

And in fact at the beginning of this century, W.M. Smart actually proved that you couldn't prove solar system stability, which is the strongest argument against Velikovsky. More recently there's a book out called *Chaos Gaia Eros* by Ralph Abraham; he had two chapters on Velikovsky and he's been taking a lot of guff. His students asked him, when he gave his lecture at Princeton, "What do you think of Velikovsky?" And he said, "Well, let me look. Is solar system stability proven?" And he came to the same conclusion that everybody else did who is an expert in the field—or nearly everyone—that you couldn't prove the solar system was absolutely stable.

GONZO SCIENCE: Abraham said that even Newton understood that the stability of the solar system was an article of faith.

GINENTHAL: Yes. He realized from Perturbation theory, that if a planet gets next to a big planet—say Mars whenever it gets close to Jupiter—it gets pulled a little further out of its orbit. They will say, "It will always go back when it's on the other side," but the concept that was developed in chaos theory with these three experts now shows that that isn't so. The math doesn't work out anymore.

But this is the great problem: we are living with certain theories that have been created in the early part of the 1900s, and these have been built up by thousands and thousands of scholars who have spent their whole life cutting their teeth getting their masters, their doctorates, doing contributions in the field, looking at the evidence from only that point of view. And now asking thousands of scholars to say—what was it one historian said to Velikovsky? "You may be right but I just can't accept it, it would undo everything I've done. It's absolutely unacceptable and I just can't face up to such a thing!"

It may not be that bad with others, but many of them, you know, really simply say, "There is absolutely no evidence to support Velikovsky, all his predictions are of no value, therefore, why pay attention?"

These people made predictions before the space age. They made tons of predictions. They said Mars would look like—of course they saw sand storms on it—they said Mars would look like the Sahara. Velikovsky said Mars was torn by great rifts in its interactions, that it would look like a planet that had undergone enormous catastrophe. When they got there, the evidence didn't fit their theory, it fit Velikovsky's theory. In fact, one of the geologists who looked at the evidence said, "Look, Velikovsky appears to be right. What it seems to me is that some time in the very early history of the solar system, Mars had an interaction with an enormous body, which created all of this."

Prediction is one of the aspects that Einstein turned to. He said, "If you predict something that nobody anticipates." And everybody said that these predictions were not possible, you know, that there was nothing in the evidence to suggest that Jupiter had a magnetic field, that Mars would be a planet torn apart with great rifts and valleys, or high mountains, that the moon would have a high thermal gradient, that Venus would be extraordinarily hot, that Venus may rotate backwards.

When you look at the totality of the predictions, and nearly all of them fitting Velikovsky's theory, then you understand the Velikovskians. Why should we give up the ghost? And we shall continue to work, and I'll continue to work—particularly to have that theory tested in space, to see if electromagnetism plays a role in space. If it does, it means Velikovsky's right, and it means that plasma physics plays an extremely important role in the nature of the universe. But it's going to take I don't know what to change things.

I don't know why theories change, or why people change, but they do. And I don't think it'll be in my lifetime, but eventually if it turns out to be correct, I only hope Velikovsky will be given the credit for his insights, rather than being called a crank or a man before his time who was really a crank who had no evidence to support his theories. Many people who are attacking Velikovsky, as before, haven't read all the literature, they haven't read one tenth of the literature. And they're still at it....

GONZO SCIENCE: Don't they say that Einstein wrote things in the margins of *Worlds in Collision* like, "Nonsense"?

GINENTHAL: Yes, he would write that, and he also wrote wonderful things, too. He wrote other things. He said to Velikovsky, "You've proven the case. You have proven the case that there were catastrophes on Earth. But not Venus…" He grabbed his head and said, "But Venus, Venus, for God's sake, you don't move planets around!"

All the astronomers and all the other people had to say was, "He has a marvelous theory, but no, no, you don't move planets around."

And that should have been enough. But what they did is they began to talk about things that they really didn't know were true. And they made statements that were not true. And these became part of the literature. And these have become part of the ethos of the scientific community.

Recommended Reading: *Dark Matter, Missing Planets and New Comets* by Tom Van Flandern; Metaresearch.org; *The Structure of Scientific Revolutions* by T.S. Kuhn; *Worlds in Collision* by Immanuel Velikovsky; *A Guide to Velikovsky's Sources* by Bob Forest; *Scientists Confront Velikovsky* ed. by Donald Goldsmith; *Scientists Confront Scientists Who Confront Velikovsky* ed. by Lewis M. Greenberg and Warner B. Sizemore; *Carl Sagan and Immanuel Velikovsky* by Charles Ginenthal; *Science Frontiers* by William Corliss

PART 4:
Electromagnetic/Plasma Anomalies

CHAPTER 10

Electromagnetic Radiation: Controversy or Conspiracy?

PAUL BRODEUR, A JOURNALIST AND AUTHOR OF *THE Great Power-Line Cover Up, The Zapping of America,* and *Currents of Death,* generated controversy in 1968 when he revealed the negative health effects of asbestos. Since then, he has chronicled the evidence that the electromagnetic pollution that is a byproduct of the modern electronic and high-tech lifestyle can contribute to a number of serious medical conditions, such as cancer.

A quick rundown of the sources of electromagnetic radiation of various kinds and strengths illustrates the potential scope of the problem. According to *The Body Electric: Electromagnetism and the Foundation of Life* by Robert O. Becker, M.D., and Gary Selden, the sources of electromagnetic radiation include the following: Every battery-operated device, home appliance, and work machine. All anti-theft systems in stores and libraries and all metal detectors. Industrial uses of strong magnetic fields. Power rails for electric trains, which radiate waves a hundred miles out from the entire length of track. All switching stations, over half a million miles of high-voltage power lines, plus "innumerable smaller lines feed[ing] into every home, office, factory, and military base." High voltage lines, which are, "in effect, giant antennae...the largest 'radio' transmitters in the world." Metal objects near power lines, which concentrate the fields to higher levels. Electromagnetic radiation produced from "air and sea navigation, time references, emergency signals, some amateur radio channels, and military communications." All radios and radio transmitters, including ham radios, police and taxi radios, more than 35 million CB radios, more than ten thousand commercial radio and TV stations, and more than "seven million other radio transmitters, not counting the millions operated by the military." Spy satellites, weather satellites, communication satellites, radar, medical machines, more than ten million microwave ovens, and more than a quarter million microwave phone and TV relay towers. "When superconducting cables are introduced, they'll increase the field strength around power lines by a factor of ten to twenty."

The Controversy Breaks

In 1976, the scientific establishment reacted with consternation to a *New Yorker* article in which Brodeur claimed that microwaves exert a profound effect on the central nervous systems of rhesus monkeys and other primates.

Ellie Adair, a leading authority on the body's temperature-regulating mechanism, and who exposed her monkeys to microwaves on a regular basis, thought Brodeur's claims were bunk. The levels of microwave exposure in her lab apparently caused no harm, and the monkeys could even control the level of exposure themselves. Adair and her husband, physicist Bob Adair, could call upon decades of open research that established the safety parameters of different forms of electromagnetic radiation.

The safety research in question was started by the Army in World War II, when a technician passing by a transmitter noticed the chocolate bar in his pocket had melted. The Army set out to evaluate any potential hazards to its personnel, and much of the research into microwaves is still supported by the Department of Defense.

Bob Adair argued that DNA damage was the mechanism known to cause cancer with ionizing radiation such as x-rays or ultraviolet (photons which can break chemical bonds are called ionizing radiation). But, in contrast to the ionized x-rays or ultraviolet, *microwave* photons can merely twist and bend chemical bonds, and cannot sever them. Therefore, according to the Adairs, microwaves are harmless.

Robert Park of the American Physical Society also disputes Brodeur's claims. He contends that the association between power lines, magnetic fields, and diseases (like cancer and leukemia—in other words, Brodeur's chief safety concern) is simply fear mongering and bad science. Park, obviously in full agreement with the Adairs, casts suspicion on Brodeur for his cold war-era journalist's mindset, ever on the lookout for cover-ups and conspiracies.

As Park put it in his *Voodoo Science*, "The lowest energy photons capable of directly breaking chemical bonds are in the near ultraviolet region the spectrum, just beyond…visible light. These photons are about a million times more energetic than the microwave photons Ellie Adair was using" (p. 148). Park states that there is no known biological response to electromagnetic fields that would "lead one to expect harmful effects" and that there are "at most a few contradictory reports of weak biological responses." Bob Adair is quoted as saying that anyone who believed electromagnetic fields could promote cancer "would believe in perpetual motion or cold fusion."

Were the federal government and the electronics industry—like Adair—downplaying legitimate safety concerns about microwaves, and thereby sabotaging the subsequent debate about power line safety, one which continues to rage? We want to know.

Park and the Adairs apparently find no evidence of the suppression of information about this public health concern. Perhaps their reasoning is that since the alleged phenomenon is not valid, evidence of government and corporate wrongdoing is simply irrelevant.

But how deep can you go into the controversy before it constitutes conspiracy theory?

Since the 1940s, the U.S. military has generated reams of research and documents that state unequivocally that electromagnetic radiation is by and large harmless. And not just harmless, but actually creates no biological effects whatsoever.

The exception is a certain threshold at which one type of electromagnetic radiation—microwaves—causes body tissues to heat up faster than the body can dissipate this heat. But all other electromagnetic radiation below this thermal level has been officially regarded as harmless. If it doesn't actually heat your skin, and most consumer electronics such as cell phones don't, possible health effects are categorically dismissed. The body has been largely supposed to be completely invisible and irrelevant to electromagnetic radiation, according to the military-industrial-scientific complex.

However, there is also 50 years of science that shows that electromagnetic radiation does indeed have biological effects below the thermal level. This flew in the face of theory in the 1940s, and according to Park it *still* flies in the face of theory.

The first studies to show non-thermal biological effects of microwaves were done in 1948 at the State University of Iowa by A.W. Richardson (no relation). Richardson and his colleagues showed that high- and low-power microwaves cause cataracts with no heating of the eye. Since microwaves can create bio-effects without heating, the door is wide open for other kinds of electromagnetic radiation to affect the body. Potentially any exposure to electromagnetic radiation—previously supposed to be innocuous—could have health effects.

Richardson's results (and the subsequent results of hundreds of scientists over 50 years into all corners of this issue) constitute genuine anomalies of science. The problem in this case, as is so often the case with inconvenient facts, is that the establishment fails to recognize a mechanism by which the observed anomalous effect could be produced. If a mechanism (or cause) is not found, then the observed effect is dissed and dismissed.

A recent study from the UK published in *New Scientist Online News* has found hard evidence of biological effects—in this case, fertility levels—from microwave levels too low to produce heat. This was supposed to be impossible, but sure enough, a nonthermal level of microwaves affects the fertility of a certain worm. This study is getting some notice because it was done in the context of cell phone safety. This high-profile issue just may help expose the myth of the total harmlessness of electromagnetic radiation, undoubtedly to the chagrin of Park and other establishment apologists.

Modern humans have painted themselves into a corner. Dependent on technology, it is no wonder that the government actively keeps the health effects secret from us.

Solving the Riddle of Bio-Effects

It's curious that Robert Park chose to focus his *Voodoo Science* attack on journalist Brodeur rather than two-time Nobel nominee Robert Becker.

Becker's career is one awesome act of pioneering research, as well as a constant uphill struggle against scientific and governmental stonewalling and bureaucracy. Unlike Brodeur, Becker's scientific credentials are as big as a house. If Park had addressed Becker's work, he would have been forced to argue his case wholly on its scientific merits.

Robert Becker is the all-time world champion in the field of biological electricity and regeneration. In 1990 he observed that despite an "overwhelming" amount of evidence indicating major bio-effects from artificial electromagnetic fields, classical physicists have been reluctant to validate the data, because they knew of no mechanism by which such effects could occur. Without such a mechanism, any laboratory or epidemiological studies indicating bio-effects will be dismissed as impossible.

But, a trail of solid facts seems to point to a mechanism.

A molecule or atom that has been disrupted by strong electromagnetic energy, resulting in the ejection of electrons from the material, is "ionized," electrically unbalanced and having a net electrical charge. Electromagnetic frequencies higher than visible light have the energy to produce chemical reactions that are damaging to cells by causing ionization.

Non-ionizing radiation was believed not to cause bio-effects except in "the gross production of shock or heat" (*Voodoo Science*, p. 231). However, studies emerging from the Navy's gigantic land-based antenna array, codenamed Project SANGUINE, indicated medically significant increased levels of serum triglyceride after a single day's exposure to non-ionizing 45 and 75 Hz. Becker and many other scientists urged further research, and expressed grave concern for the public, many of whom live in proximity to 60 Hz power lines, which fall between the two SANGUINE frequencies. The Navy's response was to deny everything and force the repatriated Nazi scientist who had done the research into retirement. (We'll detail this case in "The Science-Swastika Connection.")

Since 45, 60, and 75 Hz could not cause heat effects, clearly another mechanism was at work. Becker cites the groundbreaking work of doctors Susan Bawin and W. Ross Adey of Loma Linda University. They reported in 1976 an increase of calcium ions emerging from living nerve cells irradiated with 16 Hz fields. The experiment was replicated in other labs, which reported the same phenomenon at slightly different frequencies. This resulted in a debate over whether chance or a hidden variable was responsible for the difference.

EPA scientists Carl Blackman and Abraham Liboff of Oakland University, working independently in 1985, readdressed the attempts to duplicate Bawin and Adey's research, and integrated it with findings from England that pertained to abnormal growth rates of yeast cells under conditions of nuclear magnetic resonance. Blackman and Liboff concluded that the local steady state magnetic field of the Earth at the different laboratories was the hidden variable that accounted for the different frequencies.

By applying the equations for what's known as cyclotron resonance to the reported frequencies, and factoring in the local strengths of the Earth's magnetic field, their results accounted for the earlier findings.

In *Cross Currents,* Becker offers the following explanation of cyclotron resonance, demystifying the supposedly impossible mechanism by which non-thermal electromagnetic radiation affects biological systems:

> If a charged particle or ion is exposed to a steady magnetic field in space, it will begin to go into a circular, or orbital, motion at right angles to the applied magnetic field...[E]nergy is transferred from the electric field to the charged particle...Many of the activities of living cells involve charged particles—such as the common

ions of sodium, calcium, and potassium—acting on or passing through the cell membrane. Cyclotron resonance has the ability to transfer energy to these ions and to cause them to move more rapidly. These effects will change the function of living cells by enabling the ions to pass through the cell membranes more effectively or in greater numbers. Cyclotron resonance is a mechanism of action that enables very low strength electromagnetic fields, acting in concert with the Earth's geomagnetic field, to produce major biological effects by concentrating the energy in the applied field upon specific particles, such as the biologically important ions of sodium, calcium, potassium and lithium…The equation for cyclotron resonance says that as the strength of the steady state magnetic field decreases, the frequency of the oscillating electric or magnetic field needed to produce resonance also decreases. This is particularly significant when the average strength of the Earth's magnetic field (between 0.2 and 0.6 gauss) is put into the equation: the frequencies for the oscillating fields that are needed to produce resonance with the biologically important ions turn out to be in the ELF (extremely low frequency) region. The ELF frequencies…become the most significant part of our electromagnetic environment…Cyclotron resonance provides an understandable and valid mechanism of action for the biological effects of both normal and abnormal electromagnetic fields (p. 235-237).

In 1984, Becker proposed that the timing mechanism for cell division might be tied into the Earth's natural steady state magnetic field. Becker's writings chronicle the principles and mechanisms relating to life's relationship to the planet's magnetic field.

He also remains one of the boldest voices in identifying the potential negative bio-effects of artificial electromagnetic fields, and the inadequacy of current safety standards. Most sobering, Becker's story provides a window into the great lengths to which military and corporate interests will go to protect to their profits and their ability to exploit these effects to their own ends. These tactics include stonewalling, denying, and forcing repatriated Nazi scientists into retirement.

See more about the Becker/military-industrial connection in "The Science-Swastika Connection."

Recommended Reading: *The Body Electric* by Robert O. Becker, M.D., and Gary Selden; *Cross Currents* by Robert O. Becker, M.D.

CHAPTER 11

The Plasma Connection

WE THINK THAT PLASMA PHYSICS CAN LAUNCH A gonzo revolution in science.

Plasma physicists are largely against the dominant Big Bang cosmology, and substitute their own plasma cosmology which sees the idea of "a beginning of the universe" as an unnecessary and unscientific idea.

Plasma physics is making inroads into other disciplines as well, and promising to upset more apple carts than just cosmology. For instance, plasma physics overlaps to a significant degree the outlaw field of cold fusion, and appears to offer a leverage point where this revolutionary science might gain a handhold.

This is extremely gonzo territory.

Consider the interlaced heresies involved: Plasma physicists reject the Big Bang. Plasma physics overlaps cold fusion research. Cold fusion research can replicate ball lightning on a microscopic scale, and these are some of the same cold fusion experiments that seem to be creating new elements in a kind of alchemical reaction, which of course is scientifically *verboten*. Ball lightning (and its attendant family of "earth lights" phenomena) is probably the weird reality behind most UFO reports. Ball lightning also clearly belongs to the same family of bizarre natural mysteries as glowing tornadoes and earthquake lights.

Since the hard science of plasma physics encroaches upon all this highly weird, scientifically contentious stuff, who knows what might happen next.

The undisputed champion and father of plasma physics is Nobel Prize laureate Hannes Alfven. One of his most important contributions to science is his finding that electromagnetic phenomena—specifically the behavior of electrified gasses known as plasmas, considered to be a fourth state of matter—can be scaled up or down to infinity. As Eric Lerner writes in his book about plasma cosmology, *The Big Bang Never Happened*:

> From Maxwell's laws [Alfven] derives rules with which a researcher can develop small-scale laboratory models of large-scale astrophysical processes…He found that certain key variables do not change with scale—electrical resistance, velocity, and energy all remained the same. Other quantities do change: for example, time is scaled as size, so if a process is a million times smaller, it occurs a million times faster. Thus the stately processes of the cosmos, ranging from auroras lasting hours to [solar] prominences lasting days to galaxies lasting billions of years, can all be modeled in the lab by rapid discharges lasting millionths of a second (page 192).

Compare that to the following synopsis of various scientific anomalies from the fringes of plasma physics. From a paper by researcher Edward Lewis titled "Recent Experiments that Produced Fundamental Anomalies for Novel Hypotheses Concerning the Production of Elements, Superconductivity, and Anomalous Radiation":

> From the 1950s to the 1990s, some experimenters produced fundamentally anomalous phenomena. Bostick's electrical discharge devices produced stable dense plasma structures ("plasmoids") that existed for long periods of time and exhibited the structure and behavior of astrophysical phenomena [like emitting beams from the axis of rotation]. K. Shoulders' research showed that plasmoids and atoms interconverted and converted to electricity

and light. Matsumoto produced evidence that during electrolysis, palladium composites may convert to plasmoids . . . and may convert to new elements and light and electricity. From various evidence the plasmoids behave like ball lightning.

Edwards essentially concludes that *all* phenomena are plasmoidal in nature—matter and energy as interconverting plasma phenomena.

Certainly many phenomena are. Consider the case of the plasma vortex UFO. As reported in the May 20, 1990, edition of the London *Observer*: a research vessel from Tokyo University tracked a 400-meter-long "object" traveling north over the Pacific at a speed in excess of Mach 4. The Japanese scientists "identified the object as a plasma vortex," which they say was "caused by freak weather." This scientifically recognized phenomenon "is similar to ball lightning and believed to be generated by 'mini-tornadoes' of electrically charged air," according to the *Observer*.

The *Observer* further states that, "Plasma vortices can be luminous at night," and goes on to quote the director of the Oxford-based Tornado and Storm Research Organization, Dr. Terence Meaden: "They are often mistaken for UFOs."

Thus we see, like Shoulders' contention that atoms and plasma interconvert, some things that we might think of as solid objects could be something more like plasma instead. This not only goes for UFOs and other strange lights, but atoms and galaxies as well.

Fire Your Astronomer and Hire a Plasma Physicist

In the world of space science, conventional notions of astrophysics are top dog. These notions inform and circumscribe the basis of every scrap of research. These traditional astrophysical notions are primarily pinned on gravity and mass as the fundamental features of the universe, and every object or event in the universe can be understood through these complementary notions of mass and gravity. This view reigns unquestioned in the ranks of virtually every space scientist, from solar system experts to stargazing observational astronomers to big-picture cosmologists.

But this point of view is being seriously challenged by the field of plasma physics. Unlike the community of astrophysicists and space scientists, plasma physicists are not fed a diet of mass and gravity. The paradigm they inhabit is one of plasmas and charge. Plasma is considered to be the fundamental state of matter, and the laws that govern it, in turn govern the universe. Everything is derived from these laws, and gravity and mass are relegated to second-banana status—just like the laws of plasma physics are ghettoized in the prevailing astrophysical view.

The dominant mass-and-gravity view, which accepts a Big Bang origin without question, is pretty well entrenched but under fire from several quarters within its own ranks. For instance, one hopeful successor to the Big Bang may be found in the Quasi Steady State Cosmology (QSSC), which, we discussed in Chapter 4. Although rejecting the Big Bang premise of a beginning of time, the QSSC is still fundamentally beholden to gravity and mass as the primary constituents of cosmic organization.

But plasma physics represents a much greater threat to the Big Bang and its offshoot, the QSSC, because the plasma physicists are acting as if they have ALREADY WON. Plasma physics is its own science altogether, and every feature of the universe makes perfect sense within its boundaries. They have their own plasma journals, with their own peers reviewing their papers. They have their own conferences to attend, and everyone there intrinsically accepts the same fundamental principles. So the Big Bang is undergoing yet another spastic twitch. Who cares?

Recommended Reading: *Handbook of Unusual Natural Phenomena* and *Science Frontiers* by William R. Corliss; "Recent Experiments that Produced Fundamental Anomalies for Novel Hypotheses Concerning the Production of Elements, Superconductivity, and Anomalous Radiation," a paper posted at http://www.padrak.com/ine/ELEWIS7.html; *Space-Time Transients and Unusual Events* by Michael Persinger and Gyslaine Lafreniere; *Earth Lights Revelation* by Paul Devereux; Plasma Physicist Anthony Peratt's site at http://public.lanl.gov/alp/plasma/TheUniverse.html; *The Big Bang Never Happened* by Eric Lerner

CHAPTER 12

The Ball Lightning and UFO Mysteries Solved

The Ball Lightning Mystery

What is the weird phenomenon called ball lightning? The gonzo fields of cold fusion and plasma physics may be able to explain it.

While ball lightning is recognized by science, it eludes a full explanation. It is a luminous mass that can exist in any weather. Normally spherical, it can assume a range of shapes. It can change color, size, direction, and velocity. It can be as small as a pea or as big as a house. Ball lightning has been seen phasing through solid objects such as walls, windows, hillsides, and cliffs. It can fade silently away or violently explode. It can appear as a single ball or as a collection of lights flying in formation.

In 2001, the UK Royal Society published a collection of papers in a special edition of their journal *Philosophical Transactions* that dealt specifically with ball lightning theories. The theories were presented by a physical chemist, a physicist, and a chemical engineer, and they were allowed to comment on the other's work. This was done in the spirit of "critical confrontation," in the laudable hopes of provoking new insights via intellectual cross-pollination. This represents perhaps the most thorough treatment of the elusive ball lightning phenomenon in contemporary science, and shows how ball lightning is climbing out of its scientific backwater and into respectability.

This collection of ball lightning papers doubled the extant number of officially recognized ball lightning reports in the scientific literature, and provided three new theories about what ball lightning may be and where its energy comes from. Those theories are:

1.) Ball lightning is composed of hydrated ions/water droplets and releases energy through ion reactions. In this theory, ball lightning is an electrochemically contained plasma whose structure is maintained by a careful balance of temperature, pressure, electromagnetic fields, and gravitational fields.

2.) Ball lightning is composed of polymer threads and releases energy through surface electrical discharge. In this theory, natural particles of dust from such sources as cellulose, soot, or silica can form threads which may aggregate into a highly charged ball, which glows as it discharges gas and its surface breaks down electrically.

3.) Ball lightning is composed of metal nanoparticle chains and releases energy through surface oxidation of metal nanoparticles. In this theory, ordinary lightning strikes can cause materials like soil or wood to discharge a metallic vapor, which condenses into a ball of networked metal nanoparticles.

So the theorists are, apparently, closing in on a viable explanation. Theory #3 emerges as the strongest contender of this pack for its ability to explain how ball lightning could pass through walls and closed windows, as has been reported.

However, all of the above theories are weakened by dealing with only the most conservative definition of ball lightning: a small glowing ball that appears exclusively during thunderstorms, and only lasts about ten seconds. To have the scope that a strong theory requires, a real ball lightning theory would have to account for the sightings of anomalous ball lightning phenomena in which there was no thunderstorm present, or where the ball lasts a long time, or when the ball lightning is as big as a house. In addition, none of the theories can explain the reports of *dark* ball lightning (which may not be as strange as it sounds, since dark auroras are scientifically recognized).

In our view, a strong contender for best ball lightning explanation comes from the vicinity of the cold fusion field, which is deeply connected to plasma phenomena. Cold fusion researcher Edward Lewis shows that ball lightning and tornadoes are manifestations of the same plasmoidal phenomena. Lewis' paper, "Recent Experiments that Produced Fundamental Anomalies for Novel Hypotheses Concerning the Production of Elements, Superconductivity, and Anomalous Radiation," was given at the Second International Conference on Low Energy Nuclear Reactions.

According to Lewis, Matsumoto's cold fusion experiments created microscopic ball lightning-like plasmoids which left pits, marks, and traces in his experimental apparatus. This indicated that the microscopic plasmoids, and by extension, macroscopic ball lightning, were characteristically torroidal in nature—i.e. spinning vortices like tornadoes. Recall plasma pioneer Hannes Alfven's observation that plasma phenomena may be scaled up or down with ease.

Many reports of large-scale ball lightning in nature also reveal a torroidal structure. Both ball lightning and tornadoes leave trails, holes, and furrows in the ground, and both reportedly use hopping and skipping motions. Tornadoes have an electromagnetic aspect to them and are sometimes reported to glow and otherwise emit light and heat. Ball lightning has been seen to precede tornadoes and even to *convert into* tornadoes, and tornadoes have been seen to emit ball lightning and other forms of lightning and electricity. Huge plasmoid structures can exist in thunderclouds that then convert to electrically active tornadoes. Lewis states that "there are reports of cylindrical ball lightning sheering apart into disks, and of disks combining to form cylinders"—while some tornadoes and plasma phenomena have been described as stacks of rings or layers which move semi-independently in a rippling motion. Matsumoto's microscopic ball lightning traces show evidence of this same ring structure, which may combine into cylindrical plasmoids not that different from glowing tornadoes.

Lewis has discovered that the literature on ball lightning significantly overlaps with the reports of nature's weirdest tornadoes. He shows that intermediate phenomena link ball lightning and tornadoes, and that the two phenomena interconvert. In his words, this shows "the identity of phenomena which people have thought were disparate."

The UFO Mystery

It looks as if ball lightning and plasma physics can help solve the UFO mystery right here on our home planet. Paul Devereux went a long way toward doing this in his excellent presentation of the ball-lightning-as-UFO explanation in the book *Earth Lights Revelation*. In it, Devereux asserts that in the daylight, ball lightning appears metallic and silver, like a blob of mercury or air bubbles under water. This accounts for the otherwise difficult-to-explain "daylight disc" aspect of the UFO phenomenon, whereby metallic flying saucers are reported in the daytime.

The anomaly catalogs of William R. Corliss show that ball lightning's interesting "earth light" cousins include earthquake lights, sky flashes, expanding balls of light (or EBLs), slow low-altitude luminous "meteors," and a host of marine light displays including light bands, light wheels, expanding light rings, and whirling light crescents. One can't help but notice that Corliss' list of earth lights-style phenomena overlap with descriptions of UFOs.

Actually, Michael Persinger spearheaded the effort to formulate a physical theory of UFOs as natural phenomena, and Devereux subsequently built on this foundation by pairing Persinger's theory with even more studies, data, and ideas of his own. And according to the Persinger theory, large-scale tectonic strain-fields generate light phenomena through "piezoelectricity" in the Earth's crust. (Piezoelectricity can be demonstrated on a small scale by striking certain rocks together in the dark to produce sparks and glows. Petrified wood works nicely.)

There are now plenty of studies and data on record linking UFO activity to fault lines, seismic activity, and various sorts of electromagnetic phenomena. Based on this information, one can definitively say that UFOs are no more from space than earthquakes or the geomagnetic index.

Earth lights can also explain the alien abduction experience. After all, the power of magnetic fields to alter brain function is well documented, and Persinger demonstrated in peer-reviewed journals that waking dreams and tactile hallucinations are easily induced by applying electrical impulses to the brain. In some cases, therefore, a close encounter with something like a UFO-ish electrified plasma might also evoke incredibly realistic mental imagery, such as the classic abduction scenario. This is a reasonable explanation for ghost sightings too, and possibly some poltergeist phenomena, not to mention occasional visits by the Virgin Mary.

This "natural" theory of UFOs and the paranormal is slowly gaining credence in the scientific world. On January 22, 2002, we read a news story that originally ran in the *Anchorage Daily News* about some earthquake scientists paying a lot of attention to legends of luminous phenomena preceding earthquakes. This is the edge of the earth lights hypothesis peeking out from under the rug. It's a paradigm shift in action.

The most wondrous possibility hinted at by Devereux is that some of these light forms may be imbued with self-awareness, and actually be alive. Witnesses have described ball lightning acting "curious," "inquisitive," or seemingly "sniffing around the edges of a room" or acting playful and mischievous.

We must face the possibility that we share our planet with such life forms—aliens from our backyard.

Recommended Reading: *Handbook of Unusual Natural Phenomena* and *Science Frontiers* by William R. Corliss; "Recent Experiments that Produced Fundamental Anomalies for Novel Hypotheses Concerning the Production of Elements, Superconductivity, and Anomalous Radiation," a paper posted at http://www.padrak.com/ine/ELEWIS7.html; *Space-Time Transients and Unusual Events* by Michael Persinger and Gyslaine Lafreniere; *Earth Lights Revelation* by Paul Devereux

PART 5:
Heretical Biology

CHAPTER 13

Evolution from Space

THE FOSSIL RECORD SHOWS THAT WITHIN TWO hundred million years of the cooling of the primordial Earth, there was life, leading some biologists to speak of "instant" life.

Unfortunately for this view, the mathematical odds against the situation create a dim view of its possibility. Two hundred million years is a virtual blink of the eye in geological terms, and as we achieve better resolution of this challenging slice of the fossil record, the gap continues to close. The origin of life is receding farther and farther back toward the point of Earth's origin.

In addition, scientists are always discovering new places they never thought they'd find life. On reactor cores, inside rocks, in underwater toxic steam vents—time and again, the most inhospitable environments are found to be crawling with life. The list for what life needs in order to take hold keeps getting shorter and shorter. Some people are saying you might not even need water.

Therefore it's reasonable to ask: Did life even evolve on Earth? We're starting to think otherwise. In fact, we're starting to think that life began in deep space, drifting here as single-cell microorganisms and viruses, and then evolved as best they could on Earth. Maybe these organisms became freeze-dried from space conditions and remained dormant for billions of years. We've seen plenty of good arguments for what could keep such microspacefarers alive through the radiation they would encounter in space. These single-celled organisms could conceivably survive the entry into an atmosphere and then set up shop.

We like Terence McKenna's vision of mushroom spores percolating up into space, becoming freeze-dried and getting picked up by the solar wind, where, someday, blown to other planets, they take their chances.

The theory that life arrived here from elsewhere—perhaps

as comet-born microbes, drifting spores, or meteors bearing microscopic life—is called "Panspermia." The theory has a distinguished pedigree, originating with the Swedish chemist and Nobel laureate Svante August Arrhenius in 1907.

Despite strong support, Panspermia is heretical to most biologists and space scientists. Nevertheless, it continues to receive robust support from a vocal minority of heavy-hitters, including the famed co-discoverer of the DNA molecule, Nobel laureate Francis Crick.

The theory's most persistent modern advocates have been astronomers Chandra Wickramasinghe and the late Fred Hoyle. They chose to frame the argument partly in mathematical terms. It goes like this: to evolve biological life you must first evolve about 30 critical enzymes. The problem is, these enzymes are all terribly complex molecules. The odds of any one of them occurring by chance in the young Earth's prebiotic organic soup are astronomically slim. Plus, you need 30 of them, which only worsens the situation. The way these guys see the math, it's like having a tornado blow through a junkyard and produce a fully functional 747.

Time is on Panspermia's side, too. In the conventional theory, you'd have to wait around for billions upon billions of years for these enzymes to finally form themselves, and then they'd have to get themselves together into a living cell that can commence to evolve. And the fossil record shows these primitive cells were already going full-bore as soon as the Earth ceased to be molten about six billion years ago.

The conclusion we are left with is that life is adrift on the winds of space and just waiting to touch down. In this view, viral and bacteriological life is in fact *still* touching down—causing disease, sure, but spurring on evolution as well.

There is in fact empirical infrared evidence of biological material flung far between the stars, with spectra from space closely matching that of earthborn bacteria. In particular, low-temperature areas in molecular clouds could easily produce large molecules, and with all the time in the world, they could become cosmic hatcheries.

Wickramasinghe has been conducting some experiments that offer strong confirmation that microbes are in fact continually raining down from space. The international team that performed the experiment also features Wickramasinghe as well as Jayant Narlikar, who was a frequent collaborator with Hoyle and co-creator of the Quasi Steady State Theory. The experiments have sent up cryogenic samplers on high-altitude balloons (launched by the Indian Space Research Organisation) to collect air from the upper stratosphere, 41 kilometers up. This is virtually the edge of space. Super-strict protocols are followed to guard against the possibility of contamination from Earth microbes. The many clumps of cells found in the samples are detected using a fluorescent dye that is only absorbed by living cells. And the way the cell clumps have been found to be distributed in the atmosphere varies with height in such a way that indicates the microbes are arriving from space, as opposed to wafting up from below.

These experiments have collected the hardest evidence yet for Panspermia: actual living bacterial life forms from the edge of space. As Wickramasinghe was quoted as saying in the *Daily University Science News*, 41 kilometers is "well above the local tropopause (16 km), above which no air from lower down would normally be transported." In other words, some percolation between atmospheric layers is expected, but nothing like this.

No one except the Panspermia theorists expected that microbes would be found this high up. But these spectacular, well-designed experiments are very difficult to ignore. No matter how you slice 'em, the experiments are paradigm-busters.

The cell clumps may be from space, which would support the Panspermia theory and rewrite biology from the ground up. Or, they are native to the upper stratosphere, which itself is an outrageous idea, and will cause no small degree of harrumphing from traditional biologists and atmospheric scientists.

Recommended Reading: *Evolution From Space, Diseases From Space,* and *Living Comets* by Fred Hoyle and Chandra Wickramasinghe

CHAPTER 14

The Aquatic Ape

OUR SPECIES IS SUPPOSED TO BE SIMILAR TO OUR HOMINID cousins like the chimp, the gorilla, and so on. But we possess several biological features that they strangely lack. For example, we are largely hairless. We also have a layer of subcutaneous fat that is entirely absent in the other apes. Where did these factors come from? We need to look at the other animals that exclusively share these adaptations with us—the dolphin, the manatee, the hippopotamus, and the seal, among others.

It seems that the only reason a mammal evolves this complex of features is to become aquatic, that is to say, to go from the land into the water. Humans mostly came back out again, although it's hard to deny that in the coastal regions, there is still a symbiotic, embryonic relationship with the ocean.

The theory of an aquatic or semi-aquatic heritage for our species has a lot of explanatory power. Hair loses its insulating power in the water and becomes a liability for the swimmer or diver. A subcutaneous fat layer develops in place of hair for heat retention, and helps to fill in and streamline the body's hydrodynamics. The conscious breath control of the diving mammal is a first prerequisite for speech. As if that wasn't enough, we share several other features with the aquatic mammals. Glandular modifications allow us to weep salty tears. The infantile peach fuzz body hair we're all born with displays a hydrodynamic waterflow pattern. And upright walking would easily phase in as a response to a lifetime of wading out to the deep. In other words, it looks like we learned to swim before we could walk.

Those biologists who are against the aquatic ape theory, and who've bothered to formulate a response to it, have their work cut out for them. And of course there are alternate explanations for these physiological features.

The problem with any such laundry list of explanations is that none of them connects the dots. Each anomaly is explained away with a separate explanation or as a coincidence. All this violates the doctrine of simplicity that states, in essence, "A single explanation is better than multiple explanations." In other words, a good hypothesis unites anomalies under one roof, instead of messily farming them out.

The conventional theory—that humankind descended from the trees and spread to the African savannah—was developed in the 1800s from what little fossil evidence was available. These days, like all established theories, the "savannah hypothesis" does little to address the pesky new facts that have surfaced in the meantime.

The "old and busted" theory is that we dropped out of the trees and began roaming the grasslands. The "new hotness" theory has a million-year surf party in the middle. We dropped out of the trees only to fill the marshes, swamps, estuaries, and tidepools, prodding our big brains to develop with iodine-rich seafood. Only then, naked and diving, mouths full of abalone, did we heed the call of the dry land and swim to shore.

All right, everybody out of the pool!

The fossil record contains a likely simian candidate for our aquatic ancestor. It seems there was a swamp-living ape that evolved on an island off of Italy. And of course islands are known for producing way-out fauna. It seems this upright ape disappeared in the fossil record right as Australopithecus strode onto the scene in the African veldt. Are they one and the same? Some researchers think so.

Terence McKenna adds another wrinkle to the story of our evolution, not by supporting an aquatic ape ancestry, but by adding psychedelic mushrooms to the mix. Did early human-apes discover these cow-patty delicacies, and then start following the local wild cattle around? Looking for food, they found God, and a little linguistic stimulation....What's becoming obvious is that human prehistory was pretty gonzo.

Recommended Reading: *The Aquatic Ape Hypothesis* by Elaine Morgan; *The Archaic Revival* by Terence McKenna

CHAPTER 15

Animal Cognition

THE ANIMALS ARE FED UP. IN THE PAST FEW YEARS there've been mass animal attacks on villages across Africa and India. The baboons and the elephants are both in on this action. We remember reading one story in *Fortean Times* where some elephants broke into this wine storage vault, drank the wine, and went on a drunken rampage, uprooting signs, knocking down houses, you name it. It happens all the time now. It's blamed on "human encroachment on wild areas," but there's no doubt what this really means. Come on, elephants have an aesthetic sense and can be taught to paint; they cry when they're sad and mourn their dead; the human quality of their memory is legendary. Is it such a stretch to think human encroachment has them hopping mad?

Knowing what we know about the complexity of animal communications, we find it all too easy to imagine these elephants planning their attacks in a group huddle. What do people think, that these rampages are just coincidental synchronous attacks of multiple individual elephants? No! The elephants are tired, hungry, and consistently harassed by humans. Don't tell us they're not standing around trying to figure out what to do about bulldozers and other nuisances.

People always wonder what it'll be like when we make contact with alien intelligence, or even if we'll be able to recognize it. How about dolphins? There's alien intelligence. Did you know they can probably project sonar holograms to each other? And we're trying to teach them to sing "Happy Birthday."

The chimpanzee, often cited as our closest relative because of the near-exact match of our genetic materials, acts a lot like us, too. The male chimps gather into war parties and beat the crap out of the males in neighboring tribes. In anthropology the "chimp-as-near-human" model has had a lot of influence on how we view and define ourselves. Unfortunately, this model focuses attention on the lowest possible cognitive denominator in both species. We would argue that this has helped to cause a subtle cynicism about both species, and by extension, all animal life. The ultimate extension of this cynicism is the "ultra-Darwinist" position that organisms are just vehicles for genes, which do all the real driving.

Have you heard about the bonobo? Here's your alternative, positive model of animal cognition. Bonobos look almost exactly like chimps. But—get this—a great deal of the time they walk upright, and they have face-to-face intercourse. So human of them, hey?

But bonobos are peaceful. Instead of settling conflicts through chimpy violence, they defuse tensions using gifts of food and sex. No one has time to fight because they're always laying around, eating fruit, and getting it on missionary style. Sex only lasts a few seconds, but they do it a hundred times a day. The photo that sums it up best is this bonobo dude making overtures to a couple of bonobo chicks. He's standing there holding a piece of food in each hand and sportin' a bonobo boner. One gets a sense that these creatures are still in Eden.

According to the Gaia hypothesis, imbalances in the world system are self-regulated back into equilibrium, as if life and the environment are a single entity with a giant immune system. Our modern brand of deadly weather, as well as the increase in animal terrorism and all the other threats to our survival, can thus be understood as the advance wave of a sweeping immune system response.

Mark these words: someday the elephants and dolphins are going to come to us, and they're going to be pissed off. They're going to have a list of demands, and if it's ignored, it's going to be time for "gorilla warfare."

Recommended Reading: *When Elephants Weep* by Susan McCarthy and Jeffrey Moussaieff Mason; *Bonobo, the Hidden Ape* by Frans DeWaal

CHAPTER 16

Interview with Science Writer Deborah Blum

DEBORAH BLUM WADES INTO CONTROVERSIAL subjects without fear. Her book *Sex On the Brain* tackled the prickly subject of biologically hard-wired gender differences. In *The Monkey Wars*, she explored the debate over animal testing and examined the point where the need for it encounters the science of animal cognition. We spoke with Blum in Madison, Wisconsin, at the Association of Alternative Newsweeklies Convention in June 2002.

GONZO SCIENCE: Let's start with an irritating question perhaps, about being a woman science writer. Did you encounter any bias?

BLUM: No. You know, being a woman science writer is an interesting thing because—let me back up and say, women tend to do very little of what I think of as the "jock sciences." So if you looked at the way science gets covered, and you looked at the reporters who do hard science—what people call hard science, [namely] physics, astrophysics, astronomy—90 percent of them are men.

And so one of my biggest complaints as a science writer is that you don't see enough women covering those kinds of sciences, and all of those sciences tend to be high prestige. It's almost like, well, women cover what people now call soft science, and therefore, it's a lower level of science reporting.

But that's my own kind of gender shtick. I did not get really ly dinged for being a woman. I know some of the women scientists I talked to said that they were sometimes encouraged to pull their punches. And that they got clobbered by feminists for doing research which showed—and this is very anecdotal, because I never had a feminist say this to me—but that feminists really sort of banged on their head for doing this sort of biology of behavior stuff. But this has been a longstanding sort of position of the feminist movement, that [gender identity is] not biology but culture and environment.

I happen to not believe that. I mean, I happen to believe that biology and behavior dance together, and that neither explain all of who we are, but both explain some. And I almost mentally think of it like a dance: you're out on the floor, biology is one partner, behavior is the other, they're fox trotting around the room, sometimes one leads, sometimes the other, and I don't think we have a clue. But what I do think is that we'll never understand who we are if we pretend that there's a solo dancer out there.

So that's my position. And that part of it wasn't incredibly controversial.

The one time someone really got on my case actually had to do with a real small issue. And that was my use of the word genetic "defect." And that was because…in *Sex On the Brain*, there's a chapter in which I'm just sort of walking through the biological evidence of what is male and what is female. And my shtick is that it's a spectrum. You know, we all exist on this sort of male/female spectrum.

But, there are some definitive things. And there's a group of, there's a small genetic—now I'm going to say genetic "variation," and I'll tell you why—there's a genetic variation, and it's only in the Dominican Republic. And the name for the variation is *guevedoces*— "eggs at 12." And this is some kind of genetic glitch in which you don't get the standard testosterone surge when the body is developing, right about at birth, right on the edge of that.

And so the testes don't descend at all, and these boys look like girls. They look feminine, they have much more of a sort of feminine appearance and sexual structure, until puberty. And at 12, when you get the big jolt of testosterone, the testes descend, they go through the normal puberty then, and are clearly boys.

It's a very bizarre well-studied kind of example of hormones doing the body building, physical appearance kind of thing. And so I wrote about that, and in the book I called it a genetic defect.

And I ended up like on a transgender radio station at MIT. And they asked me—it was actually after I did this sort of summary of the book in the *Utne Reader*—and they called me up and said, "We read that, we liked the story, we'd like you to be on the show," and then they sort of came at me.

GONZO SCIENCE: They set a trap.

BLUM: Yeah, basically. It didn't bother me that it was a trap, because I think that's part of the game, once you're out there talking about your work.

I'm really defensive about getting things wrong. If I make a mistake, I beat myself up, and then if someone jumps me with a mistake, I just hate it. But if it's something where, you know, they're talking about a particular issue—they're not saying "You got this wrong," but they're saying, "We don't like the way you expressed this"—then what I really want to do is think through what they had to say. So that didn't bother me so much.

And at first what I wanted to do was argue about it. You know, all scientists say "genetic defect"; that's what they say. That's just the standard lingo, and so here they are jumping on my case—

GONZO SCIENCE: As if you had an agenda in saying that.

BLUM: Right. This woman said, "Well, what you're basically saying is these people are defective." And we actually ended up in kind of a standoff.

I never really conceded the point on the show, but I went back and thought about it, and I thought, "Yeah, you know, I can see that." I can see that if you were someone who was living with a— now I'm going to say genetic variation—and people were continually describing you as having a defect, then you might filter that out as being defective. And I'm not interested in doing that. I'm trying to make a point about how biology works; I'm really not trying to slap people around.

And so now I say genetic variation. And you know, maybe it's a little P.C., but I don't care. Genetic variation is just a little less loaded. And so that's what I'll say.

The trick about this is that I don't know if people will know what I mean. If I say this is a defect, people get it just like that. If I say this is a genetic variation, that's not a very standard way to describe it. And so now I've introduced a weird little bit of ambiguity. So now I have to add in something that says, "a genetic variation in which the gene does not do what it does in the usual sense," or something like that. So it makes it a little clunkier.

GONZO SCIENCE: Do you feel you have a responsibility as a science writer to change the language, or the way the lingo is used, to inform the issue?

BLUM: That's a good question. Do I think it's my responsibility? I think it's my responsibility to be respectful. People are mad at me all the time...I've spent my whole career making people angry, because I like gray areas and I like controversial issues. I actually made fewer people angry with that sex book than lots of other things that I've done.

GONZO SCIENCE: Like *The Monkey Wars*?

BLUM: Yeah, *Monkey Wars* made people angry. I mean, there were people who wouldn't speak to me after that book.

GONZO SCIENCE: Did you make animal rights people angry?

BLUM: Scientists.

GONZO SCIENCE: Scientists, really?

BLUM: Yeah. Animal rights people just didn't love that book, because it didn't say that animal research is bad. I mean basically, it didn't take a stand at all. In fact if it had taken a stand, I probably would have made more money. I didn't make that much money, because when I wrote that book, I wanted to do two things. The sort of basic theme of the book is that animal research is about us. It's about the number one species on the planet, right, that can do whatever it wants with another species, even a closely related, really smart species. We still call the shots.

And so every decision we make about animal research says something about who we are, our ethics, our morals, how we use the animals, how we respond to them, how we treat them, how we deal with each other over those ethical issues.

And so the whole book, every chapter except one, is built around a person's story, which is sort of the narrative device; it's a series of stories. And one starts with the death of a monkey. And every other one starts differently than that. I don't say anything in that book about whether I think animal research is good or bad. I say we have moral and ethical responsibilities here.

And I had people come up to me afterwards and say, "Well, I could really tell what you thought." And I would think to myself, "Well that's interesting," because I wasn't sure what I thought, you know?

In the end, I think that there are no good alternatives for all of animal research; we haven't got them yet. We would love to replace animal research with computer simulation, and with cell culture, and with all of the other little tricks in which you don't engage a whole-body system. But we can't do that well yet; we don't understand how the body works well enough to accurately simulate it. We don't understand the species-to-species differences, right?

So I think that in terms of the priorities . . . if our first priority is to make sure that we continue to try to improve medicine and deal with health problems, then we can't get away from animal research yet. And I say "yet" because I think the long-term goal should be not doing [animal research]. But we're not there. And as long as we aren't there, then I think there's lots of moral and ethical responsibilities to do it right, do it fair, do it decent, treat the animals with respect.

And do I think that scientists engage in all those opportunities? They don't. There are a lot of scientists who do, but science is not a monolithic kind of one-person enterprise, and there are lots of scientists who are very resistant.

GONZO SCIENCE: You must have heard about this discovery that some tribes of chimps are using stone-age technology.

BLUM: Yeah, and then that chimps have culture—different hand-shakes.

GONZO SCIENCE: Do you think that this is going to inform this issue?

BLUM: People try. I think that someone in New Zealand introduced civil rights legislation for chimpanzees. It didn't pass, but it was introduced. And there's a really interesting book by a Boston-area lawyer called *Rattling the Cage* in which he looks at the intelligence, capabilities, and social functioning of chimpanzees, and makes the argument that we should grant civil rights to chimpanzees.

GONZO SCIENCE: Carl Sagan was almost saying that in *Dragons of Eden*. He asked, "Why are chimpanzees in jail?"

BLUM: If you spend any time with chimpanzees, [you'll find that] they are so enormously smart, and they comprehend so much, that it really can get painful after a while. And yet, we've known chim-panzees are that smart for at least two decades.

GONZO SCIENCE: Or should have known.

BLUM: And we're just getting people to deal with that. I think that the big fear in the science community is that chimpanzees are the thin end of the wedge. Once you give that to another species, then what next?

I was at an animal intelligence and cognition meeting in Chicago, it must have been a couple summers ago, and they were talking about the chimpanzee cultures in Africa—you know, with their different handshakes, and their different hellos—it's really amazing stuff. But there are some guys who can show some of the same differences in whale cultures.

And one of the things I do think we do—that's very hard to

get past in our uneasy relationship with our own biology—is we really want to believe we're separate from all these other species. We're writing about these other species but we're not one of them, so, we're above that. And religions tend to preach that. And certainly the Christian religion—you know, this huge justification for doing anything you wanted, which I find very annoying. I think that we can make a very strong case that chimpanzees deserve some kind of legal protection they're not getting. And I think that I will not hold my breath waiting for that to happen. It'll never happen in *this* administration [Bush II].

And here's the torque of that: we've got all these chimpanzees in captivity in this country, and some of them—at least the ones that have outlived their research potential—are basically in jail. They can live for 50 years. You know, they're just locked up. There's no doubt about that. Those chimpanzees are treated really badly.

Not all of them. If you went out, say, to the Yerkes Regional Primate Research Colony in Atlanta, they have a field colony of chimpanzees. That's what they do: they observe chimpanzee social interactions. And they have some separately caged, but they have a huge colony that's all about social give and take. So you get some of that, too—and this doesn't justify it—but we don't treat chimpanzees well anywhere. We don't treat them well in captivity, and we don't treat them well in the wild.

And at this meeting about chimpanzee culture, one of the really interesting threads that kept coming was Japanese whaling, and the bush meat crisis in Africa, where the companies that are cutting their way into the forest—to feed the workers, they give them guns and encourage them to go ahead and shoot the animals. And there are a lot of chimpanzees being shot to death. For food. And Jane Goodall actually came to that meeting and talked about how bad this was.

You know, two wrongs don't make a right, as we all hear when we're growing up, but the fact is that it's very consistent. The way we treat animals in research—and we do it much better in this country than we used to—is consistent with the way we treat animals. We're not particularly respectful of other species.

CHAPTER 17

Toward a Field Theory of Biology

MATTER USED TO BE WHAT YOU COULD KICK. BUT AT the quantum level, what we call particles of matter are actually knots of excitation in the surrounding energy fields. Matter is now understood as a state of energy, and the billiard-ball physics of Newton is no longer sufficient. The deeper reality is the ubiquitous energy field of space-time, which generates, shapes, and sustains all particles. This field theory is needed to describe matter at this level.

On a related note, a reductionist conceptual error dominates the science of biology. Species and organisms are no longer considered to be the ones who mutate and evolve—only the genes are supposed to do this. The mainstream Darwinists (currently referring to themselves as neo-Darwinists) have all of evolution and biology explained through the gene alone, and its enshrined constituent molecule, DNA. Sloppy, harder-to-quantify organisms have disappeared from the neo-Darwinists' view screen in favor of a cleaner, nuts-and-bolts molecular chemistry.

Like the physics of Newton, this mechanistic approach is not sufficient. One often-overlooked reason for this is that, as biologist Brian Goodwin wrote in *How the Leopard Changed Its Spots*, "molecular composition does not in general determine form." Consequently, the biological puzzles of form—morphology, embryology, and regeneration—cannot be explained purely through the action of the genes.

To understand these puzzles it is helpful to begin with blastema cells. A blastema is a mass of undifferentiated cells produced at an injury site in a regenerating animal, such as a salamander. The blastema eventually specializes into nerve, muscle, and whatever else is needed.

Let's say our salamander has lost its leg. Blastema cells—primitive cells uncommitted as yet to specific functions—form on the wound. To grow into the appropriate leg cells, all genes for all other cell types must be repressed.

How do the blastema cells know which genes to repress? Looked at through the reductive lens of molecular chemistry, atom "A" in molecule "B" somehow knows that it must help form a leg instead of a tail. Not only that, but it must know where to form into bone, or flesh, or nerve. The situation is the same in embryology: What tells the DNA how to express itself?

What's missing here in this mechanistic model of molecular building blocks is any notion of a principle of spatial organization. Ultimately, there must be a feedback loop to an organizing force acting within the context of the whole organism.

As demanded by the above considerations, and by quantum physics itself, we must dispense with the mechanistic fallacy. We must admit that organisms are dynamic fields of particular kinds. Neo-Darwinism ignores the spatial patterns and dimensions of the organism. To describe those, you need a field theory.

Welcome to the morphogenetic fields—resonant, vibratory fields that organize the dynamic interplay, and relational order, of the molecular components of life.

What characteristics can these m-fields be said to have? Can they be quantified?

Two-time Nobel nominee Robert Becker, M.D., maintains that an m-field corresponds to the bio-electric field. Becker's ideas are substantially testable and grounded in hard science, and he's had some astounding successes in their application to the medical field.

He proved that at least some of the acupuncture points fall precisely upon nodes of electromagnetic potential on the skin (which belies the comment in the *Skeptical Inquirer* a few years ago that no reputable scientist believes in acupuncture).

Becker is also the guy who stimulated partial regeneration in rats by electrically stimulating a blastema to form at an amputation site. Regeneration in mammals is totally unheard of, but, thanks to Becker, someday in the not-so-distant future, we might be able to re-grow our lost limbs.

GONZO SCIENCE

Biologist Rupert Sheldrake conceives of m-fields primarily as structures of probability. Sheldrake's hypothesis invokes a non-energetic capacity for the transfer of information across time and space called morphic resonance. In Sheldrake's view, all systems and structures resonate with their own pasts and those that are similar. This self-resonance is cumulative and results in m-fields having a built-in memory. Testing for phenomena based on the effects of "morphic resonance" has been the basis of his two most recent books, *Seven Experiments that Could Change the World* and *Dogs that Know When their Owners Are ComingHome*. While his hypothesis is controversial even among those scientists closest to him, it is the Gonzo opinion that he's certainly gotten a hold of something. It's always a delight to watch the skeptics try to explain away his ever-growing body of evidence.

One conclusion that Sheldrake comes to is that the "laws of nature" are more like "habits." To back this up he cites evidence that the speed of light may actually be in flux, and of course it seems like every time they measure the gravitational constant it's different.

Brian Goodwin finds Rupert Sheldrake's recourse to a non-physical interpretation of morphogenetic fields to be unnecessary when addressing the shortcomings of mainstream biology that they both acknowledge. He argues that neo-Darwinism is an incomplete theory which suffers because of its myopic focus on the molecular components or organisms. He strives to bring biology closer to physics by emphasizing the principles of organization that come from the science of complexity. He is a radical because he recognizes that organisms are a distinct level of biological order and as such are as irreducible as the molecules that comprise them. In Goodwin's work, morphogenetic fields refer to "spatial organizing activities that involve clearly defined physical and chemical processes, though combined in a way that is distinctive to the living state." Like Becker and Sheldrake, the implications of Goodwin's work have far-reaching implications for the future of science. According to *How the Leopard Changed Its Spots:*

Inheritance and natural selection continue to play significant roles in this expanded biology, but they become parts of a more comprehensive dynamical theory of life that is focused on the dynamics of emergent properties.

The consequences of this altered perspective are considerable, particularly in relation to the status of organisms, their creative potential, and the qualities of life. Organisms cease to be mere survival machines and assume intrinsic value, having worth in and of themselves, like works of art. Such a realization arises from an altered understanding of the nature of organisms as center of autonomous action and creativity, connected with a causal agency that cannot be described as mechanical. It is relational order among components that matters more than material composition in living processes, so that emergent qualities predominate over quantities. This consequence extends to social structure, where relationships, creativity, and values are of primary significance. As a result, values enter fundamentally into the appreciation of the nature of life, and biology takes on the properties of a science of qualities. This is not in conflict with the predominant science of quantities, but it does have a different focus and emphasis (p. xii).

The science of qualities is being put into practice in the Holistic Science program now being taught at Schumacher College in Devon, U.K.

Scientists such as Brian Goodwin, Robert Becker and

Rupert Sheldrake are an incredible threat to the status quo—witness the book reviewer in *Nature* who suggested that Sheldrake's books be burned. Read them while you can.

Recommended Reading: www.gn.apc.org./schumacher college/; *How the Leopard Changed Its Spots* by Brian Goodwin; *Signs of Life* by Goodwin and Richard Solé; *Form and Transformation* by Goodwin and Gerry Webster, anything by Rupert Sheldrake; *Body Electric* by Robert Becker with Gary Selden.

CHAPTER 18

The Human Genome Sucks

WHAT THE NIMRODS DECODING THE HUMAN GENOME failed to understand is that genes are just parts. If a disease is a disease of the whole, then tinkering with the parts is only going to work for a limited time.

Let's say for argument's sake that you have an immune disorder like cancer or arthritis. There is an overwhelming environmental component that taxes the entire immune system. These kinds of diseases are on the rise, and it doesn't take a genius to see that the chemicals in our environment contribute to them.

So, what, the genome guys are going to go in there and tweak your brain cancer gene and make it all go away? At best you'll get a temporary fix, because the environmental disease agents are ever-present and streaming through your tissues. It's like if you eat sand all day, and they try to solve your digestive problems by giving you a new colon. But hello, what about all the sand?

The promise of new miracle cures stems from the same conceptual error. They've just opened the world's biggest erector set, and all they can think of is to build the same old boners they've been screwing us with for years. This is the mechanistic fallacy, folks. This is billiard-ball physics all the way to ultra-Darwinism, where genes do all the real evolving and the organism is reduced to a vessel.

What the human genome posse might be able to do for you is to work up some shots that you'll have to take for the rest of your life. Basically, this is just constantly re-fixing the problem, since nothing is going to keep the environment away from you. What's worrisome is that re-fixing a chronic problem only amounts to suppressing the symptoms.

What we're left with is that the cures are bullshit and they may even kill you but maybe they're still better than the disease. The uncomfortable choice is your own; this dilemma may have only one horn after all.

Is it too much to ask that they quit thinking of new ways to make the same old mistakes? Take genetically modified food. Will shooting virus genes into my corn improve my life so much that it's going to outweigh the incalculable (and unknown) risks of trans-species genetic pollution?

This human genome/genetic engineering nonsense is, fearsomely, drowning out the quite rational voices from the organics movement who don't want genetically modified everything, with genetically modified pollen cross-contaminating entire fields and flushing family-farm organic livelihoods down the drain when organic certifications get revoked due to loss of species identity.

It's already happening. Organic shipments have been turned away at the checkpoint because the corn chips have salmon genes—not to mention the whole Starlink fiasco, and the pharmacological soybeans in the corn fiasco, and all the other fiascos and political maneuvering that spell contaminated food supply. The safeguards in place to protect us from foreseeable genetic disasters seem eerily familiar to the nuclear industry's sterling safety record. From where we're sitting, the genetic modification of food crops has been a debacle. Genetic this and genetic that. Don't believe the hype.

Meanwhile, a funny thing happened on the way to the human genome. They thought—no, they knew—that there were 100,000 genes to be mapped. This grand project sheepishly finished early and we learned that well, there are only 30,000 genes. Fantastically, every scientist involved was dumbfounded, and battle lines were drastically redrawn overnight. They were in all the papers saying things like, golly, mice have only 10,000 genes—how can we have only three times as many genes as a mouse? Why, humans are not even far ahead of yeast!

But even that insult paled in comparison to the discovery that humans actually have fewer genes than rice.

It's like our manhood was riding on this count. Bigger is better, right? Do we have to mention that this really shortchanges mice, yeast, and rice? These things are animate matter, after all, and last time we checked, that was still pretty special. Must we be diminished in our comparison to them?

CHAPTER 19

Interview with Systems Theorist Fritjof Capra

FRITJOF CAPRA IS A PHYSICIST, WORLD-RENOWNED author, and faculty member of Schumaker College. Since the publication of his influential book *The Tao of Physics* 20 years ago, he's moved more toward the life sciences and ecology. In that vein, he has become the founding director of the Berkeley-based Center for Ecoliteracy, which promotes ecology and "systems thinking" in primary and secondary education.

At the time of this writing, Capra's most recent book is *The Science of Life: Integrating the Hidden Connections Among the Biological, Cognitive, and Social Dimensions of Life*. The phrase "hidden connections" of the title is derived from Czech statesman Vaclav Havel, who said, "Education today is the ability to perceive the hidden connections between phenomena."

Capra expounded upon this quote while giving the keynote address at the Fourth International Conference on Science and Consciousness in April 2002. Capra explained:

> In science, we refer to this ability [to perceive hidden connections] as systemic thinking, or systems thinking. It means thinking in terms of relationships, connections, patterns; in terms of context…complexity, and complexity theory.… I extend the systems approach…to the social arena, to social phenomena, and I apply it to some of the major issues of our time, such as globalization, biotechnology, the whole question of sustainability, ecology, ecological design, and so on.

The Tao of Physics applied systemic thinking to the question of where Eastern mysticism and Western science overlap. The book laid bare the connections between the world of quantum physics and the "all is one" mindset of the mystics. It made Capra a virtual hero in some quarters, and he subsequently became the de facto spokesperson for a new breed of holistically minded scientists.

The book also made him a target. Some scientists (in particular the hard-nosed professional-skeptic crowd) didn't like the way Capra's book inspired a legion of "New Age" copycats. The skeptics grew increasingly aggravated by the loosey-goosey "scientific mysticism" which sprang up in its wake.

Capra's work has since grown to encompass many diverse scientific issues. Since constrictive modes and styles of analysis—such as mechanism and reductionism—have long obfuscated many scientific issues, it would seem that a new approach is needed. The lens of systemic thinking is well suited to filling this need; it can clarify and sharpen the debates occurring within any field.

On the other hand, mechanism and reductionism each represent the incisive power, as well as the crippling blind spots, of modern science. When a field is gripped by these dual trends, it races to narrowly define its central questions in terms of whatever can be measured in a test tube, reducing each issue to its smallest, most isolatable components. While this has led to great accomplishments, it has also obscured the forest for the trees.

Biology is a prime example of this situation. Heavily influenced by atomic physics, modern biology has lost sight of the organism in a way, choosing instead to focus on molecular chemistry. Biologists have been so used to thinking in terms of genes and molecules that they have left themselves open to being astonished by recent discoveries that belie this simple view.

We spoke with Fritjof Capra in Albuquerque at the Fourth International Conference on Science and Consciousness in April 2002. Before discussing issues of biology, we asked him to answer his detractors from the world of professional skepticism.

GONZO SCIENCE: We're looking at *How To Think About Weird Things* by Lewis Schick and Theodore Vaughn, Jr., both card-carrying members of CSICOP. They say, "Capra can't claim that modern physics vindicates the worldview of Eastern mystics in general, because the Eastern mystics don't share a common worldview." How would you refute that?

CAPRA: Well, this is a very superficial critique. Although the Eastern mystics have great diversity of views, they share certain common concepts about the world, about the nature of reality. And these common concepts are the ones that I'm referring to in the work I did about 20 years ago in *The Tao of Physics*. And I do quote mystics from various schools, from traditions of Hinduism, of Buddhism, of Taoism, and I've also gone on to use Christian mystics. It is well known among scholars today that mysticism is a religious tradition that is global, and is based on basic religious experience which is then interpreted differently in different cultures and different traditions. But they all have a lot in common.

So you can compare scientific theories with these philosophies. You can also say, of course, that in physics, and in biology and psychology and Western science, there are lots of theories. If you ask scientists today about the theory of evolution, for instance—this is maybe a good example—you will get lots of different views, but no one will dispute evolution as a fact, that all species evolved from common ancestors. What people are arguing about is the detailed dynamics of this process of evolution. But they all accept the common framework. So the creationists, for instance, deny that common framework. And similarly you could say those skeptics you just mentioned are ignorant of the common framework of mystical traditions.

GONZO SCIENCE: What are your thoughts on the Human Genome Project, and whether their results meant what they thought they would?

CAPRA: Well, this is very interesting, and in my new book I have a long chapter on genetics and biotechnology. And what the Human Genome Project—which was and is a tremendous achievement—has shown is the necessity to go beyond genetic determinism, this philosophical stance that says genes determine behavior.

And although it's true—it is true that a genetic basis is an important aspect of every biological process and behavior—genes do not determine behavior in a simplistic, linear way. Genes are part of a whole cellular network that has multiple feedback loops. And we need to study this, the so-called "epigenetic network," to even understand the functioning of genes. Even the genome itself is a network where different genes influence other genes.

And so the Human Genome Project has now triggered a conceptual revolution in biology, which is extremely fascinating. And it also calls into question the basic attitude of biotechnologists, who say, "Once we have understood the coding of genes, and the maps of genes on the genome—on the DNA—we will understand the origin of disease, of human behavior, and so on."

That's just nonsense, because you have to know more than just the structure of the genes. So the shift now is a shift from the structure of DNA to the principles of organization of biological networks. It's an extremely interesting development.

GONZO SCIENCE: So would you say that the work of non-mechanistic biologists Brian Goodwin and Rupert Sheldrake have sort of put their fingers on what establishment biology has perhaps missed?

CAPRA: Well, I wouldn't put them in one basket; I wouldn't mention them in one breath, Goodwin and Sheldrake. They present quite different views.

Goodwin, like myself, is a systemic thinker and uses a systemic approach to biology, talking about networks and emergent properties. He uses complexity theory in biology. Sheldrake is what I would call...a neo-vitalist.

In their critique, they are similar. They say, "The mechanistic approach to biology doesn't get the whole story; we need something else." But then they differ in saying what that something else is. Sheldrake says that there's a non-material force—or field, a morphic field—that is an entity that needs to be understood and taken into account.

Goodwin will say it's not a separate, non-material entity that we need to understand, it's a set of relationships among the components, and we need to move from just objects to relationships, to understanding the patterns and relationships and processes. So they are quite different in their approaches. But both agree in their critique that the mechanistic approach doesn't work.

Recommended Reading: Anything by Rupert Sheldrake, Brian Goodwin, and Fritjof Capra.

CHAPTER 20

Interview with Biologist Rupert Sheldrake

WE SPOKE WITH RUPERT SHELDRAKE IN ALBUQUERQUE, at the Fourth International Conference on Science and Consciousness in April 2002. He is a key player in the debate about the nature of biology, as well as an influential writer, thinker, and critic concerning the politics and the practice of science.

GONZO SCIENCE: When did you begin to think that the mechanistic explanation for biological phenomena was not sufficient?

SHELDRAKE: I think [it happened] when I was an undergraduate at Cambridge. I began to see that it was much too limited, that there was so much about animals and plants that you just wouldn't understand in terms of chemistry, anatomy, and physiology and so forth.

I think I'd first become alienated before I went to Cambridge. I worked in a research lab in a pharmaceutical company…And without intending it, I ended up working in a vivisection laboratory where they were doing really heavy experiments with animals.

I'd gone into biology because I loved animals, and I kept pets as a kid. And here I was at age 17 and 18, and I worked at a lab where animals were routinely butchered, vivisected, poisoned. And it made me feel there was something terribly wrong in the way I was going in biology, because that's not what I thought biology was about. I thought it was about studying living things, not tormenting them and killing them.

I think the emotional impact of that experience made me think more deeply about things than I otherwise might have done. I came to the conclusion that mechanistic biology was all right as far as it goes, but most of things I found most interesting were poorly explained mechanistically.

GONZO SCIENCE: Why do we need field theories to describe biological phenomena?

SHELDRAKE: We need field theories to describe biological phenomena because biological phenomena seem to be field-like. We need field theories to describe, for example, magnets and the iron filings around it, because well, you can describe it without one, but you can't understand them without one. And when [electromagnetism pioneer Michael] Faraday first invented the idea of magnetic fields, it was because there were phenomena that seemed to need a field explanation.

First of all, when embryos develop…the form they take up, the shape they fall into, the shape of my body or your body or any animal body, depends not just on the genes and the chemicals, but the way they're arranged. The field plays the role of the kind of architectural plan of the organism.

If you analyze your arm or your leg, they have exactly the same genes, the same chemicals, the same bone cells, the same muscle cells. There's nothing in your arm or your leg that makes it an arm or a leg because of the chemicals in it; it's the way they're arranged. And the DNA can't explain that by itself. It's the same in all the cells and it's the same in the arm or the leg, the genetic material.

It's a bit like having two houses in a suburban street, where they're built using the same kinds of materials, but they're different shapes, because they have different architectural plans. And since the 1920s, a number of developmental biologists who study the development of form have been convinced that fields are necessary to explain this process. It's actually quite a mainstream idea in developmental biology. That's one area in which I think you need fields.

The second [reason why a field theory is needed is because] understanding the way in which groups of animals behave together— flocks of birds and schools of fish—they respond in relation to each other extremely rapidly, much more quickly than they could if they just went with their neighbors. They all seem to be responding to a field of the group.

Thirdly, I think you need fields to understand mental activity. I think these fields help to organize the activity of the brain and connect animals and people with their cubs, and with each other.

GONZO SCIENCE: Where did the resonance concept come in; when did you start to develop your hypothesis of "formative causation"?

SHELDRAKE: I first started doing it around 1972, when I was working on morphogenetic fields in developing organisms . . . I realized that if these fields exist in organisms then they have to evolve, because organisms evolve; their shapes change over time. All different kinds of animals and parts evolved from simpler ones. So the fields that organize them can't be a fixed reality; they must evolve along with the organisms.

So, how could a field evolve? In physics we don't have the idea of evolving fields, usually; the idea is that they're always the same. That couldn't really be the case with biological fields because all organisms change over time, and all species. So if fields began to evolve, how would the field be remembered? The simplest thing would be if the fields themselves had memory within them. The idea of "morphic resonance" is a memory principle, whereby past organisms influence present ones, on the basis of similarity. And this means that each species has a kind of in-built collective memory.

GONZO SCIENCE: And this has led to your current work, which we know is controversial, in which your research suggests that there is a phenomenon—through a mechanism that we're speaking of as morphic resonance—whereby animals, dogs for example, might know when their owners are coming home.

SHELDRAKE: Well, here what I'm saying is that it's not so much a memory principle—although memory plays a part in sustaining the field—but the dog and the owner, when they form an emotional attachment, they become closely bonded. I think a part of the field is the social field, which links members of social groups together. So here's the dog, there's the owner—there's a field as it were like a bubble around them, which is the field of their social relationship. Dogs bond to people quite strongly; they are after all descended from wolves, which are very social animals.

And what I'm saying is that when the person and the dog go apart—say the person goes away to work and leaves the dog at home—the field that connects them isn't broken, but it does stretch. It's like it's dumbbell-shaped, and it does stretch out just like an invisible elastic band joining the person and the dog. So when the person decides to come home, for example, the dog can pick up that intention.

And we've shown in hundreds of videotaped experiments that dogs and some cats do indeed anticipate when their owners are coming home, in a way that can't be explained in terms of routine, familiar car sounds, and so on. It seems to be essentially telepathic. And the means of communication is this bond, the field.

GONZO SCIENCE: And do the skeptics continue to discount the videotapes, all the data that you've amassed? Do they concede that there is something going on?

SHELDRAKE: Well, I mean, if we're talking about skeptics who are open-minded, then yes, people who've seen the data can see there's a huge amount of evidence there. A couple of skeptics did some experiments themselves with dogs, and they came up with almost exactly the same pattern of data as I did myself.

But if one's dealing with people who are not so much skeptics but dogmatists, I mean people who believe in their own truth and everything else must be false....

GONZO SCIENCE: Like CSICOP, for example.

SHELDRAKE: That is an organization of skeptics, some of whom within it are more open-minded than others. But dogmatic skeptics are sort of scientific fundamentalists; nothing will change the way they think. And if they're so convinced they know the truth—that telepathy and things are impossible—as far as they're concerned, *any* apparent evidence for it must be a result of fraud, deception, delusion, et cetera; it can't possibly be real.

There's nothing much one can do with people like that; but fortunately they're in a small minority within the scientific realm, and within the larger world, they're a *very* small minority. And most people are interested in the evidence.

And I think of course it's much more scientific to do experiments and look at evidence than to know in advance that something can't happen.

GONZO SCIENCE: Are you familiar with the work of Robert Becker, the orthopedic surgeon who did a lot of work with salamanders and electromagnetic fields?

SHELDRAKE: Yes. I think what Becker says is that to allow organisms to regenerate, electric currents can have an effect. He showed a range of electromagnetic effects on animals. I think that his work is important in highlighting the importance of electromagnetism in animal physiology. I don't think he's explained the shape into which things grow. Passing electric currents into a regenerating limb stump, of a frog or something, may enable it to regenerate when it wouldn't have done so otherwise.

But newts, which are very closely related to frogs, can regenerate legs without any electrical currents being passed at all. It's a natural ability; for some reason the frog can't seem to do it spontaneously. It's as if passing electric current actually removes a blockage, rather than actually underlying the process.

GONZO SCIENCE: The label "vitalist" or "neo-vitalist" has been pointed at you. I believe it was you who pointed out that many of the more neo-Darwinist folks put all this emphasis on the gene, a so-called "genetic program," that doesn't have any more explanatory power than the morphogenetic fields that you're talking about.

SHELDRAKE: We need to take into account both genes and fields. But I think the standard view in biology is dangerously close to a form of "molecular vitalism."

Vitalism is the belief that there is some mysterious principle at work within living organisms . . . I think the molecular vitalists, like Richard Dawkins and other neo-Darwinists, try and cram all the vital factors of living organisms into the genes. They say, "It's very small; it's okay to have all these mysterious vital factors."

The way they talk about genes goes way beyond what molecules, chemical molecules, can actually do. Dawkins speaks about genes being selfish; molding organisms; competing with each other; engaging in evolutionary arms races; being as ruthless as Chicago gangsters. His writings are riddled with anthropomorphic, metaphorical language. In fact, practically all his writing is metaphor. Now there's nothing wrong with metaphor; we use metaphors to understand things. But where it goes wrong is when metaphor is confused with reality; when "selfish," vitalistic genes are confused with the reality of DNA molecules. In fact the "selfish genes" are just a projection; they're projections of images and metaphors onto molecules. . . .

I think what genes do is code for particular proteins, and some genes are concerned with control of protein synthesis, but they don't explain the organism any more than the tarmac on highways in the city or concrete in the buildings explain the life of the city; they're part of living creatures. They're not even instruction books, they're just strings of nucleotide bases which say what amino acids should be strung together in what order.

GONZO SCIENCE: So the DNA of an organism is not self-replicating?

SHELDRAKE: No, the DNA is self-replicating in the sense that one strand enables the other strand to be specified, but it can only replicate of course with the whole of the rest of the organism.

GONZO SCIENCE: So, we're paraphrasing Brian Goodwin here, but if it's true that organisms have a dense spectrum of states, if there isn't a limitless number of forms that organisms could take or that the living state is a dynamic state and is subject to the laws of physics, that this is what's lost when he says and you say that the spatio-temporal elements of biological form have been overlooked.

SHELDRAKE: I think they've been overlooked because the drive has been to reduce things to the smallest possible bits. You know, it's like trying to understand a TV set by grinding it up and analyzing the silicon and copper and stuff. You find out something about it, but you destroy all the structural matrix of functions on which it depends. And you also neglect to see that the functioning of the TV set depends not just on the components inside it, but on the programs to which it's tuned, which are invisible influences, and not found by looking at its material components.

So I think it's naïve, but it may be a necessary stage for biology to go through. Molecules are certainly there. But they don't explain the whole organism and what's going on within it.

I think we're going to have a better way of understanding life, and a better system of medicine, for example, that can include enough of holistic and complementary therapies, instead of just pretending they don't exist. We're going to have to have a wider theory of life and that's going to have to be a field theory.

Recommended Reading: *The Sense of Being Stared At, and Other Aspects of the Extended Mind; The Presence of the Past;* and *A New Science of Life: The Hyporthesis of Morphic Resonance* by Rupert Sheldrake. He also operates the website www.sheldrake.org.

PART 6:
Controversial Archaeology

CHAPTER 21

The Strange and Terrible Story of the Kensington Runestone

THE COMFORTABLE SCIENTIFIC AND SCHOLARLY worlds of history, archaeology, runology, and Scandinavian linguistics have all been rocked by developments surrounding a single stone in west-central Minnesota known as the Kensington Runestone (KRS). Long derided as a hoax, to all appearances the KRS has been authenticated as a sign of a medieval Scandinavian presence in the middle of North America. This defies the current preconceptions of history, in which Columbus discovered America in 1492.

The century-long brouhaha surrounding the KRS provides a laboratory for exploring the excruciating difficulty that often attends the acceptance of new ideas in science. It has long been acknowledged that the Vikings maintained a short-lived presence in Newfoundland around the year 1000; what is so different about acknowledging medieval Scandinavians in Minnesota in 1362? We're about to find out.

The Story of the Stone

In 1898, just outside of Kensington, Minnesota, farmer Olof Ohman found an enigmatic artifact ensnarled in the roots of a tree—a large stone tablet, weighing over 200 pounds, covered with medieval Scandinavian runes.

The inscription on the stone, which now resides in a museum in Alexandria, Minnesota, reads:

8 Goths and 22 Northmen are on this acquisition expedition from Vinland far to the West. We had traps by 2 shelters one day's travel north from this stone. We were fishing one day. After we came home I found 10 men red with blood and dead. Ave Maria deliver from evils. I have 10 men by the inland sea to look after our ship, 14 days' journey from this island Year 1362.

Many features of the Kensington Runestone were so far outside the experience of the experts of the day that it was quickly branded a hoax. Ohman's life was never the same. His reputation was destroyed, and he and his family endured a virtual reign of terror that, arguably, helped drive his daughter to suicide. And the KRS itself, like Ohman's reputation, has remained an object of derision ever since.

Now, more than a hundred years later, a piece of linguistic scholarship by Richard Nielsen seems to demonstrate conclusively that the KRS inscription is genuine. Simultaneously, Nielsen's work has been joined by long-overdue scientific testing on the surface of the KRS, coordinated by chemist and design engineer Barry Hanson. As a result, the stone's authenticity now stands verified well beyond the doubts of the past.

The scientific world must suddenly grapple with an uncomfortable certainty: that medieval Europeans traveled to the middle of the North American continent more than a century before Columbus first imagined its shores. More importantly, the reputation of its discoverer, Olof Ohman, has finally been cleared, but only after the terrible human cost that was exacted upon him and his family. In a 1995 videotaped interview with Ione and Einar Bakke, Ohman's last surviving son is quoted by family friends as saying that discovering the KRS was the worst thing that ever happened to his family.

The authorship of the stone has been falsely attributed to Olof Ohman for a hundred years. A long line of scholars examined the stone's inscription and saw the unfamiliar runes, grammatical structures, and numerical notations, and declared the stone a fake, and a poor one at that.

However, the unknown numbers, letters, words, and sentence structures began to turn up in authenticated Scandinavian texts almost immediately after the stone's discovery. Indeed, several "unknown" runes had turned up even *before* the stone's discovery, yet went completely unrecognized by the top scholars of the day. In rejecting the stone, they were really only demonstrating the limits of their own knowledge.

We think that the story of the Kensington Runestone, America's oldest written document, will go down in the annals of linguistic shame.

How It Happened

Olof Ohman was a man with 36 weeks of education, nine children, and a marginal farm. According to researcher Barry Hanson, in *Kensington Runestone: A Defense of Olof Ohman, the Accused Forger,* "Olof Ohman's character, after endless scrutiny by people who were trying as hard as they could to show the KRS as a modern forgery, survives unblemished. Every person who knew Ohman said the same thing; he was honest, he was honorable and he would not or could not have carved the inscription, nor would he lie to his sons as part of any conceivable activity he might be engaged in." And yet this unlikely hoax suspect was accused of perpetrating a brazen fraud.

The case against the authenticity of the stone was primarily made using linguistic arguments. It's not too far off the mark to suggest that the way the linguists handled the case of the KRS provides a study of how not to approach a scientific anomaly. The geological analyses of the stone have always supported its medieval pedigree, with the geologists in question always adopting a "wait and see" attitude toward the disputed runes. As we now know, this was apparently the correct approach.

Thirty-two experts in Scandinavian linguistics have declared the KRS inscription fraudulent over the past hundred years. Only ten of these experts actually published papers and/or books about the KRS. None of these scholars, not even one, caught on to the fact that the stone was legitimate. One factor that helps explain their collective failure is that most of them hailed from the small, closed, conservative world of European linguistics. When seen from this stodgy milieu, the KRS was nothing short of an outrageous affront.

In addition, at the time the stone was found, the linguistic study of medieval runology was actually a fairly young science, and in many ways, woefully incomplete. Looking back on it now, this is obvious, as the past hundred years have seen developments in the field that, perhaps, could not have been anticipated back then. For instance, old manuscripts have been continually uncovered, yielding previously unrecognized medieval runeforms, words, and elements of grammatical style. There are still piles of medieval writings that present-day scholars have not even been able to analyze yet, and we should be wise to expect even more surprises.

The KRS contains 23 different runic letters (or "runeforms"), which are used to write an inscription of 46 words and seven numbers. Eleven of the runeforms, and more than a dozen of the words, were unknown to the experts of 1898. In addition, the numbers on the stone failed to conform to the proper notation conventions as they were then known. For these and other reasons, time and time again, the case was made for fraud. Once the first couple of investigators had proclaimed the KRS to be a forgery, it seems as if the taint of scandal became so intoxicating to the academic community that the stone couldn't get a fair hearing.

Needless to say, the authentication of the KRS has been an unscientific process. The way the KRS, and Olof Ohman, became objects of scorn says more about the psychological opposition to fresh ideas than the proper conduct of science. The way some of the experts comported themselves in the face of the unknown did a disservice to the ideals of the scientific method. Personal attacks took the place of data collection. Sloppy scholarship and unreferenceable claims became the order of the day. Invoking "the experts" took the place of doing actual research.

The general attitude towards the KRS was one of contempt, as evidenced by Professor Jon Helgason's remark to KRS "debunker" Erik Moltke in 1951: "In my opinion the inscription on the Kensington Stone is such that no philologist with any self respect could in any decency write about it." But rather than thrusting his nose up in the air and ignoring it, as Helgason deemed to be the proper response, Moltke expressed a more aggressive opinion as to how one should approach the anomaly. As he wrote in the same 1951 article, "there has been so much fuss made about this inscription that a stop must be put to it."

Sloppiness and hubris pervade the quality of scholarship that the stone has been subjected to through the years. In his zeal to "put a stop to it," Moltke resorted to some extremely heavy-handed rhetoric, as when he wrote of the KRS in 1949, "see what an abortion it uses as an *a*-rune." In addition, Moltke claimed that the *n*-rune had gone out of use by 1100 and that was reason enough to declare the KRS a forgery. In fact the *n* runeform was used well into the fourteenth century, as documented by various scholars since J.E. Liljegren found one in 1832. When this information became known, both Moltke and another skeptic, Sven Birger Frederik Jansson, quietly dropped it from their list of complaints. As for the *a* runeform, it is found on the Lye Church inscription from medieval times on the island of Gotland.

So many of the unknown KRS runes, flatly declared to be "impossible" in 1898, could have been found with a little research. The runes in question hadn't made it to the runic dictionaries yet, but they each existed in referenced works, and in principle, they could have been found. Simply put, the KRS was never seriously studied by those most qualified to so. It was merely written off, when in reality, it had much to teach. It is hardly an exaggeration to suggest that if the KRS had been subjected to serious scrutiny, and authenticated in 1898, it would have advanced our knowledge by a hundred years in the fields of Scandinavian linguistics, runology, history, and the archaeology of North America. As it stands, some doors in these fields are only now beginning to open, yet we've had the keys for four generations.

A case in point: one of the prominent Scandinavian linguistics texts cited in the KRS controversy was *Old Swedish Grammar with Inclusion of Old Gotlandic*, published by Adolf Noreen in 1904.

Erik Moltke, for one, heavily utilized Noreen's text to "debunk" the stone in a series of articles in various scholarly journals. As Einar Haugen and Thomas L. Markey put it, whereas Noreen's book was "a monumental achievement for its day . . . [it] has not been reprinted since it first appeared and, as many texts not readily accessible to Noreen and his readers have been published subsequently, it is badly in need of revision." Of course, fitting with the overall pattern of the KRS story, Moltke became one of the prime "experts" invoked by others as they dismissed the stone, even though he had used Noreen's outdated linguistic text from 1904 as the basis for his own conclusions made as late as the 1950s.

Many of the claims made against the inscription on the KRS appeared in the guise of unreferenceable claims, which often take the form of something like, "This word is impossible for the fourteenth century." Stated authoritatively by an "expert," these kinds of claims appeared to carry weight, and inevitably became cited by others. This is why Richard Nielsen's paper authenticating the KRS is so significant: it represents the only peer-reviewed comprehensive paper on the language of the KRS. Each claim is referenced, and each bit of data has been verified by the reviewers.

When it was discovered that Ohman owned a couple of books that contained runes, the skeptics went wild. Here was the smoking gun. The books, published in Stockholm in the late 1800s, were *The Well-Informed Schoolmaster* by Carl Rosander and *Sweden's History From the Earliest Time to Our Time* by Oscar Montelius. It was claimed that Ohman had used these books as the source material for his hoax. However, as is the case with so many other arguments leveled at Ohman and the KRS, the accusation that he used these books was never seriously investigated, but merely thrown onto the pile of already-existing accusations. If the issue had been seriously looked into, it could have been easily exposed as impossible. For the runes available to Ohman from the Rosander and Montelius books account for only half of the runes on the KRS. And of the remaining runes, many of them were unknown to anyone in 1898—not just to the uneducated farmer, but to the most educated Scandinavian linguists in the world.

But Wasn't There a Death Bed Confession?

Then there is the case of the so-called "death bed confession." While it has done much to prejudice the public against the stone's authenticity, it is quickly revealed to be the only real hoax of the entire KRS saga.

It seems that in the late 1800s there were some neighbors to the north of the Ohman farm by the name of the Gran family. In 1967, Theodore Blegen, one of the arch-nemeses of the KRS, persuaded a nephew of Walter Gran to do an audiotaped interview which became known as the "Gran tape."

Walter Gran was four years old when the KRS was unearthed, so he didn't remember much about it. But he did manage to spin a tale during the interview about how his father, John Gran, made a "death bed confession" in 1927. The "confession" allegedly concerned Olof Ohman forging the KRS.

The story was published in 1976 in a series of articles in *Minnesota History* by Russell Fridley, director of the Minnesota Historical Society. The BBC even got in on the act when they used the "confession" as part of a documentary about the KRS "hoax." The trouble is, the spectacular death bed confession is untrue in all ways.

John Gran was not on his death bed when he allegedly uttered these things. Walter Gran said that the "confession" happened in 1927, but John Gran did not actually die until 1933, six years later. Some death bed.

There is not even a confession. Walter alleged on the tape that his father said, "Go ask Ohman." Then when Walter did so, Ohman is alleged to have said nothing.

Compounding the issue, Walter Gran's truthfulness was not given glowing endorsements by his friends and neighbors. One of them said in a 1979 interview with Ted Stoa of Fargo, North Dakota, "I didn't take too much stock in what Walter said at times." This same sentiment was echoed by Iona and Einar Bakke in their 1995 videotaped interview with Ove Pederson.

Olof Ohman's character witnesses, on the other hand, never had a bad thing to say.

Vindication

As discussed, the "unknown" runes, words, grammatical quirks, and numerical notations of the KRS have all turned up in medieval Scandinavian texts. To settle the issue, all of the pieces needed to be assembled. This finally happened in a single 75-page paper based on 12 years of research. Published online by the peer-reviewed journal *Scandinavian Studies*, this slam-dunk article by Richard Nielsen destroyed the case against Ohman and the KRS. The shock waves are just beginning to reverberate.

Sixty-six years after his death, Olof Ohman has, apparently, been cleared. It's only logical, of course. If, in 1898, you assembled every expert in the world, and put them in a library full of the world's existing literature on medieval linguistics and runology, they could not have generated the full inscription on the Kensington Runestone. And yet Ohman, with virtually no education and only a couple of general reference books on his bookshelf, stood accused of this very deed. As stated by KRS researcher Barry Hanson, author of *Kensington Runestone: A Defense of Olof Ohman, the Accused Forger*, "they accused him of doing something that they couldn't even explain."

Hanson himself has played no small role in the authentication of the KRS, complementing Nielsen's linguistic work with some striking scientific studies based on geological concerns. The geologists looked at the stone and said, in effect, "This thing's ancient; there's no way it could be a modern forgery." This argument has been around for a long time. Geological considerations have supported the stone's great antiquity since N.H. Winchell, state geologist of Minnesota, and others (including the state geologist for Wisconsin) first examined the KRS in the early 1900s.

Since that time, however, the linguists have dominated the debate and successfully marginalized the issue of the stone's geological features. That is, until Hanson came along and wondered: what has been discovered upon examining the stone with a modern microscope? To his amazement, he learned that no one had ever attempted it. For that matter, no one had ever so much as *recommended* that this investigation be conducted.

Hanson realized that this represented a huge gap in the history of the KRS controversy, and got to work. First, in 2001 he published his recommendations for physical testing of the KRS in the peer-reviewed history journal *Journal of the West* as "The Kensington Runestone: Physical Features, Past and Present." Then, having been granted exclusive authority (by the owners of the KRS) to coordinate the scientific testing of the stone, Hanson contacted various people in the fields of geology, chemistry, and geophysics. Work was initiated at American Petrographic Services (APS) in St. Paul and continued at the University of Minnesota Department of Geophysics, where an electron microprobe analysis was conducted on parts of the surface of the stone. APS oversaw some scanning electron microscope work at Iowa State University on the same samples. The initial results of these investigations indicate that the original geological assessments of the KRS by Newton Winchell in 1909 are correct. In other words, the KRS is authentic, and it had been in the ground many, many years before Olof Ohman moved onto his land near Kensington.

The Kensington Runestone and *Erik's Chronicle*

Richard Nielsen, the man who finally cracked the case of the Kensington Runestone, did it in part with the help of a fourteenth century tale of knights in shining armor known as *Erik's Chronicle* (or "EK," an abbreviation of the Old Swedish title, *Erikskronikan*).

Nielsen was well on his way to authenticating the long-dismissed language of the KRS after he made a singular observation. As he wrote in "Compliance of the KRS Language With That Shown in *Erikskronikan*," "No investigator has attempted to identify an Old Swedish document that best reflects the language of the KRS. EK seemed like a good candidate owing to its length of 4560 lines that perhaps contain over 25,000 words."

The seemingly massive job of studying the 19 manuscripts that comprise EK was made somewhat easier by the extensive studies that had already been done. A scholar named G.E. Klemming had translated EK over a period of three years in the late 1860s. Then it caught the attention of Rolf Pipping, who used EK for his doctoral thesis in 1919. That same year, Pipping went on to publish an index of every single word in EK, complete with line numbers. As Nielsen wrote, "This makes EK extremely handy to study."

Contrary to the Scandinavian language experts (who have dismissed the "impossible" language of the KRS up to the present day), the great majority of the linguistic features of the KRS can be authenticated from this single fourteenth century tale. In fact, Nielsen writes, "Differences between KRS and EK are slight and not many, and the KRS in each case is directly confirmed from other [medieval Swedish] sources…"

Therefore, the KRS controversy need not have raged past 1919, when anyone involved could have picked up Pipping's index. That would have meant a lot to Olof Ohman, the poor farmer who unearthed the KRS. When Ohman died in 1935, Pipping's index had existed for 16 years, yet Ohman still had not shaken his reputation as the forger of the KRS. Pipping's EK index would have belied the accusations against Ohman, namely that he fabricated this or that word, rune, or piece of syntax.

By way of examples, here are some of the linguistic features of the Kensington Runestone that have been cited as evidence of a forgery, and how they have been laid to rest.

Case management: The supposedly incorrect use of case in the KRS is often cited as proof that the inscription was the result of not just forgery, but a somewhat unsophisticated forgery. However, Nielsen has brought to light, via EK, that the issue of case in this period of the Middle Ages is more complex. He explains:

> The KRS is written in Old Swedish, which during the 13th century underwent a rapid readjustment of its complicated case system. This case system had to manage four declensions…In addition there were three genders…in both singular and plural. This means there were 24 different cases in all…What makes the 13th century so interesting to study is its record of dissolution of these cases.

Opponents of the KRS like to take a classical
stance with no dissolution of case, but this is not
the situation that faced a scribe in 1362 (p. 3).

Essentially, the KRS experts have failed to recognize the difference between the most formal style of Old Swedish, used for, say, legal documents, and the language as it was actually used day to day.

Singular verbs in plural function: The fact that singular verbs in the KRS inscription were used with plural nouns has been pointed out among the skeptics as proof of a clumsy forgery, for surely no actual speaker of Old Swedish would fall prey to this confusion. However, it is not a case of confusion at all but merely that, like case management, the plainspoken language of 1362 differed from its most formal written counterpart (the version known by most KRS skeptics). As Nielsen writes, after citing 16 examples from EK alone, "The singular verbs in plural function on the KRS are in keeping with the spoken language of 1362 and certainly can not be considered to be proofs of forgery."

Umlauts: This is the one that started it all. Professor George Curme, one of the first people to examine the inscription, rejected its authenticity based on the fact that it contained double dots over some runes, which he construed to be umlauts. Since umlauts were not invented until the sixteenth century, he decided the stone was a fake. However, double dots in the fourteenth century were not umlauts at all but indicated vowel lengthening or the insertion of a letter. As Nielsen says, "Clearly, Prof. Curme got the investigation off on the wrong foot" (Nielsen, "Response to . . ." *Scandinavian Studies*, p. 47).

Nielsen shows how easy it would have been to verify the KRS, and to redeem Olof Ohman while he was still alive.

The Implications of an
Authenticated Kensington Runestone

What new avenues of discovery have been blown open by this scientific blockbuster?

Whoever the carver was, he was part of an "acquisition expedition" of eight Goths and 22 Northmen from the year 1362, smack dab in the middle of the North American continent. The Goths would have been from what is now western Sweden, and the Northmen could have been from anywhere else in Scandinavia.

In his article in the *Journal of the West*, Barry Hanson, building upon previous work by Richard Nielsen, speculates that the origin of the expedition may actually be related to the strange disappearance of a settlement in Greenland over 600 years ago:

>...it is known that the Western settlement in Greenland had a bad series of winters starting in 1308...In 1341, the settlement was discovered gone by Ivar Bardson, the bishop at Gardar in the Eastern settlement. There was no sign of violence or devastation; there were even some stray cattle roaming around. Some 1,500 people simply left in their boats, with many of their possessions. No one knows where they went, but it is suspected that these same people regularly visited the Ungava Bay area for wood, caribou, and fish. They also probably were familiar with the Hudson Bay area, because there is evidence that they were at the Chesterfield inlet for iron and other parts of the Bay for polar bears and eider down. Based upon the types of fur they are known to have, it is strongly suspected that these Greenlanders traded with the natives of the region...[T]ravel up the Hayes or Nelson rivers would be quite possible for the "Greenlanders."

We asked him to speculate further about the people who carved the KRS. Hanson responded:

My guess…[is that] they were from mainland
Europe but associated somehow with the
Greenlanders that had migrated from the abandoned Western colony…There most likely was an
existing population of [the Greenlanders] in the
area of the KRS. Other artifacts including two
boat hulls have been reported in this area, at 1370
[feet] elevation…1370 [feet] is the same elevation
as remnant shore erosion features at KRS hill
[water levels used to be higher in the area and
KRS hill used to be an island]. I think they took
their time doing the stone. The KRS hill maybe
was the "home base."

Hanson thinks that other artifacts will be found in the vicinity of KRS hill and possibly a few miles to the east.

Is there any other evidence that supports this new view of
early European penetration into North America? In fact, in addition
to the reports of the ancient boat hulls mentioned above, there is a
plethora of hitherto unacknowledged artifacts that may soon be getting a second look.

For instance, there are the triangular holes of the Whetstone
Valley in South Dakota and many more in western Minnesota.
These consist of hundreds of unexplained triangular holes, five to
seven inches deep, in large rocks all across western Minnesota and
northeast South Dakota. They do not appear to be blasting holes
made by pioneers. Could they be mooring holes?

There are also the so-called Chippewa Valley axes, period-
ically found in virgin soil. Mostly in the hands of private collectors,
these axe heads could represent medieval-era, hand-forged, pre-cru-
cible steel, and have no known counterparts in any mainstream
American museum. At any rate, they have never been properly
identified or studied.

There are also what appear to be habitation sites, discovered
two to three feet underground via the remote sensing emitted
infrared technology developed by Marion Dahm of Chokio,
Minnesota.

And lastly, of course, there are about half a dozen other runestones. They are all smaller and less well known than the KRS. But after what happened to Olof Ohman, is it any wonder that the discoverers of these other stones might have chosen silence, instead of scrutiny?

Why the Kensington Runestone Was Rejected by the Establishment

In the classic book *The Art of Scientific Investigation*, first published in 1950, author W.I.B. Beveridge describes many of the elements seen in the KRS affair in a chapter called "Difficulties." In this chapter, Beveridge anticipates T.S. Kuhn's treatment, some decades later, of the painful birth of new paradigms.

Beveridge may as well be speaking of the KRS controversy when he wrote in the 1957 edition of his book:

> There is in all of us a psychological tendency to resist new ideas…just as there is a psychological resistance to really radical innovations in behavior or dress. It perhaps has its origin in that inborn impulse which used to be spoken of as the herd instinct. This so-called instinct drives man to conform within certain limits to conventional customs and to oppose any considerable deviation from prevailing behavior or ideas by other members of the herd. Conversely, it gives widely held beliefs a spurious validity irrespective of whether or not they are founded on any real evidence (p. 146).

Beveridge goes on to quote Wilfred Trotter, who wrote in his *Collected Papers*: "The mind likes a strange idea as little as the body likes a strange protein and resists it with similar energy. It would not perhaps be too fanciful to say that a new idea is the most quickly acting antigen known to science. If we watch ourselves honestly we shall often find that we have begun to argue against a new idea even before it has been completely stated."

Through these quotes one can clearly discern the emerging contours of the Kensington Runestone.

Recall the conversation between Professor Jon Helgason and Eric Moltke, in which Helgason claimed of the KRS that "no philologist of any self-respect could in any decency write about it," and Moltke's considered response that, instead, "a stop must be put to it." The following words from Beveridge seem to address their reactions: "When adults first become conscious of something new they usually either attack or try to escape from it. This is called the 'attack-escape' reaction. Attack includes such mild forms as ridicule, and escape includes merely putting out of mind."

Helgason exemplifies the escape reaction by having the "decency" to never say anything in print about the KRS; presumably this means putting it out of his mind as well. Moltke, on the other hand, embodies the attack reaction, trying to kill the new ideas represented by the KRS, by "putting a stop to it." He appears to conform precisely to Beveridge's model, even to the point of stooping to ridicule, as when he compared the *a*-rune of the KRS to "an abortion."

Another aspect of the KRS controversy is touched on by Beveridge in a quote he pulls from an essay by F.C.S. Schiller, "Scientific Discovery and Logical Proof," published in *Studies in the History and Method of Science*: "One curious result of this [mental resistance to new ideas], which deserves to rank among the fundamental 'laws' of nature, is that when a discovery has finally won tardy recognition it is usually found to have been anticipated, often with cogent reasons and in great detail."

In the case of the KRS, as discussed above, there are the many instances of previously known, studied, and referenced words and runeforms that were dismissed as impossible when they were found on the Kensington Runestone. In accordance with Beveridge's analysis, the language of the KRS can thus be said to have been anticipated, and studied in great detail, long before the discovery of the actual stone and its subsequent rejection.

In light of the fact that the Ohman family suffered so terribly as a result of stumbling across the KRS artifact, it is tempting to condemn the many individuals who prematurely rushed to judge the stone. These are the people who sealed the Ohmans' fate. Beveridge is quick to sound a note of caution, however, before we turn the tables on the skeptics and heap upon them the same ridicule that they themselves have dispensed. The rejection and "debunking" which plagues such discoveries as the KRS can be understood as part of an inevitable, tragic process. According to Beveridge:

> In nearly all matters the human mind has a strong tendency to judge in the light of its own experience, knowledge and prejudices rather than on the evidence presented. Thus new ideas are judged in the light of prevailing beliefs. If the ideas are too revolutionary, that is to say, if they depart too far from reigning theories and cannot be fitted into the current body of knowledge, they will not be acceptable. When discoveries are made before their time they are almost certain to be ignored or meet with opposition which is too strong to be overcome, so in most instances they may as well not have been made. Dr. Marjory Stephenson likens discoveries made in advance of their time to long salients in warfare by which a position may be captured. If, however, the main army is too far behind to give necessary support, the advance post is lost and has to be re-taken at a later date (p. 144-145).

We are left with the disquieting conclusion that the harassment visited upon Olof Ohman and his family, at the height of which his daughter committed suicide, was the natural result of a quirk in the practice of science.

If there is a moral to the strange and terrible story of the Kensington Runestone, it may be Beveridge's admonishment that, "What we must aim at is honest, objective judgment of the evidence, freeing our minds as much as possible from opinion not based on fact, and suspend judgment where the evidence is incomplete. There is a very important distinction between a critical attitude of mind (or critical 'faculty') and a skeptical attitude."

Recommended Reading: Richard Nielsen's paper, "Response to Dr. James Knirk's Essay on the Kensington Runestone," is the first comprehensive treatment of the language of the stone to be peer-reviewed. It may be found, along with many supplemental papers, in Barry Hanson's *Kensington Runestone: A Defense of Olof Ohman, the Accused Forger*. Hanson's non-profit archaeology organization's website is Archaeologyitm.com, where one may find his book.

CHAPTER 22

The "Berg-AVM Runestone" Fiasco

BARRY HANSON'S BOOK *KENSINGTON RUNESTONE: A Defense of Olof Ohman* contains the following description of what he terms "The Berg-AVM Runestone fiasco":

> In late 1994 Mr. Bob Berg…and several of his cohorts found a stone near the Kensington Runestone Park which had a short inscription carved into it. The group could see that the letters "AVM" were chiseled into the stone as well as a date "1363"…This incident was later reported to their Viking Research group in April 1995 by Mr. Berg…[It] was reported as an obvious hoax and no-one ever thought any different until May of 2001 when a woman…and her father came along and found the same stone. They claimed it was an authentic medieval artifact and then proceeded to call a press conference without the benefit of a scientific report or having someone knowledgeable in runic inscriptions look at it. It was…promoted as authentic…the Minneapolis *Star-Tribune*…reported the incident [and] incorrectly stated that there was a connection between the so called "AVM stone" and the Kensington Runestone scientific [authentication] effort (Appendix I, p. 1).

In a privately circulated memo dated November 5, 2001, Hanson detailed the features of the Berg-AVM Runestone that gave it away as a hoax: "Obvious red flags are the fact that the Berg-AVM stone was found well below the 1,370 foot elevation level (so that it would have been underwater in 1363), it was also noted that the pyrite weathering features indicate a recent date since pyrite oxidizes quite rapidly to soluble ferrous sulfate. Indicators of great age which ARE present on the [Kensington Runestone] are not present on the Berg-AVM stone."

However, the woman who was promoting the authenticity of the AVM stone did not know any of this until it was too late. As Hanson explains in *A Defense of Olof Ohman*, the very people who could have helped her were not allowed to. "Bob Berg and his associates were prevented from even inspecting the [AVM] stone after it was 'found' as was this author [Hanson]...Lessons from the Berg-AVM fiasco might [include]...try not to exclude people who might be helpful to the investigation...Bob Berg and others were ignored or intentionally excluded from the process."

The results were predictable, because the Berg-AVM stone was soon revealed as a hoax. According to Hanson, in September 2001 two former University of Minnesota graduate students signed a letter claiming that they were part of a group of five students who committed the hoax in 1985.

"The five students were in a graduate level class, taught at the University of Minnesota, in which they determined that the nearby Kensington [Runestone] was a fraud. Their idea was to demonstrate how gullible and naïve people were who thought the [Kensington Runestone] was authentic," Hanson wrote.

The Minneapolis *Star-Tribune* story of November 6, 2001, quoted one of the Berg-AVM stone hoaxers, Kari Ellen Gade. Now a professor of Germanic studies at Indiana University, she was asked about the basis of her belief that the Kensington Runestone is a hoax like the one she committed. "All serious scholarship has drawn that conclusion," Gade said.

But as Hanson noted in his memo, the *Star-Tribune* has not done the Kensington Runestone research community any favors with its propensity to: "quot[e] people as experts [on the Kensington Runestone] because of their job description and without establishing the fact that they are indeed authoritative. In fact several of the past interviewees [about the Kensington Runestone] could not support their opinions with referenceable data. The reporter of course has failed in the past to ask for evidence or data which would support the 'expert' opinion, they have simply printed it."

This did not satisfy Hanson, whose book painstakingly examines every word ever written about the Kensington Runestone by the so-called experts.

Hanson suspected that Professor Gade—like the other "experts"—was unaware of recent developments in the study of the language of the Kensington Runestone. Issues like its authentic, rare "e" dialect have led to a revolution in understanding the artifact.

Hanson contacted Professor Gade. He informed her of the recent linguistic developments and asked her some questions about the language of the Kensington Runestone, such as: if no expert in the world knew of the "e" dialect when the Kensington Runestone was discovered, how would an alleged hoaxer be able to get it on the stone?

She refused to answer.

Gonzo Science also contacted Professor Gade. She failed to respond.

Is it possible that, as a professor of Germanic studies at a major university, Gade's grad-student role in an irresponsible hoax was causing her some discomfort at work? Or was she beginning to realize that her conclusions about the Kensington Runestone were on shaky ground?

Professor Gade is lying low.

Recommended Reading: Richard Nielsen's paper, "Response to Dr. James Knirk's Essay on the Kensington Runestone," is the first comprehensive treatment of the language of the stone to be peer-reviewed. It may be found, along with many supplemental papers, in Barry Hanson's *Kensington Runestone: A Defense of Olof Ohman, the Accused Forger.* Hanson's non-profit archaeology organization's website is Archaeologyitm.com, where one may find his book.

CHAPTER 23

Interview with Kensington Runestone Researcher Michael Zalar

MICHAEL ZALAR IS AN INDEPENDENT RESEARCHER who published an article in the peer-reviewed history journal *The NEARA Journal* that detailed literally dozens of mistakes made regarding the Kensington Runestone by the Smithsonian Institution. Zalar's article led the Smithsonian to make changes in its book *Vikings: The North Atlantic Saga*. We interviewed him in Duluth, Minnesota, in November 2001.

GONZO SCIENCE: So, how did you get involved in the Kensington Runestone controversy?

ZALAR: Well, I picked up a book at a rummage sale. It was [Robert] Hall's *The Kensington Runestone Is Genuine.* It presented a pretty positive case for the Runestone. And I went, "Okay, what's the other side of this?" And, you know, I started poking around, and things just built. I got into discussions on the Internet, which got me to researching down at the [Minnesota] Historical Society, and looking through all their stuff and building up my files, and I kind of got hit by a research bug. I love digging into stuff and finding things out, and there's plenty of stuff there.

GONZO SCIENCE: The *NEARA Journal* article detailed 40 errors about the Kensington Runestone in a book by the Smithsonian Institution.

ZALAR: Thirty-seven errors. I went for factual errors, not matters of interpretation. Things where the information that I had conflicted with what they were trying to say, with what they were saying in the book. And when you stop and think that these 37 errors occurred over about three pages of text, that's quite a bit of misinformation that was being put out.

Now, I have heard back. Brigitta Wallace, who wrote that section for the Smithsonian, did write a counter-claim, but for the most part, she does not rebut my charges. She puts out extraneous information which does not deal with what I was talking about, the very specific things that I had said.

There are a couple of places where she adds some information, like personal letters from a relative in Scandinavia, which I don't have access to, which do apparently counter my information—which does not mean that I was wrong; it merely means that she had another source for those. And that's fine, that's legitimate. If she's got a reputable source, then that's fine.

But when she misquotes published reports, changes the conclusions of a scientific inquiry 180 degrees from what was published—she said it was totally different—that's just bad scholarship. That's not something that should go in print.

GONZO SCIENCE: Have you had any correspondence or dealings with any of the other Runestone skeptics, like [prominent Runestone critic James] Knirk?

ZALAR: Yeah, I've exchanged E-mails with Dr. Knirk, who was very cordial. His assumption is that, based on the runes and the linguistics, the Kensington Runestone is false. I put forward other arguments—the history, the geochemical analysis that has been going on—and he did say that the geochemical analysis by American Petrographics and the University of Minnesota was some pretty solid science. I doubt that he'll be changing his mind on this, but he is at least open, far more open to receive new information, it would appear to me.

As a matter of fact, he's been trying to say that there's never been double-dotting over any of the runes, in manuscript form of the alphabet in medieval Scandinavia. And I was able to point out on a website a private manuscript collection—the Shoyin manuscripts—I was able to point out a very specific twelfth century instance where there was double-dotting over some of the manu-script. And he has acknowledged that. I pointed that out to Dr. [Richard] Nielsen as well, and he said that the double-dotting was used for abbreviation. I pointed that out to Dr. Knirk and he admitted that it did exist, but he didn't think it was for the purposes of abbreviation.

GONZO SCIENCE: Do you think he was surprised that it existed at all?

ZALAR: It's hard to say, from E-mails, whether there was much sur-prise. I think that maybe it was something he had not recognized, that there was double-dotting occurring in manuscript forms. But again, he seemed perfectly willing to accept that evidence, so. . .

GONZO SCIENCE: So he's a good scientist.

ZALAR: Yeah, he's one of the guys that I think you can say has some reason to disbelieve in the Kensington Runestone, and, you know, I don't think he's going to change his mind unless some very, very dramatic evidence comes forward. He's always going to challenge Dr. Nielsen, and so on, but if that evidence comes forward—like if there is evidence that this could not have taken place any later than the year 1800, I'm sure he would accept that, and realize that, yeah, there's a problem there. You can't dismiss it quite so easily.

GONZO SCIENCE: Have you ever dealt with [Runestone skeptic Kirsten] Seaver?

ZALAR: I wrote a letter to the editor of *Mercator's World* regarding an article Seaver had written. And she did respond to that letter in print. She did so in a very scholarly way. So far, I have nothing against Seaver; as a matter of fact, she's made some points which I think tend to lend themselves toward the credence of the Runestone.

For instance, she pointed out that she's found archaeolog-ical evidence which corresponds to a passage in a medieval book called the *Inventio Fortunate* which was written in the 1360s. It was about a trip that was undertaken by this anonymous monk/scholar, apparently out in the areas around Greenland. And unfortunately, the book is lost, so we can't refer to the *Inventio Fortunate*, though several other writers have referred to it; cartographers have referred to it. This would have to have been in connection to the Kensington Runestone expedition; it took place at the same time. So if there was someone out there describing these lands during the 1360s, and if the Kensington Runestone is real, there must be a connection there.

Now, she mentions that she found some archaeological evidence on Baffin Island which seems to agree with something of the bits and pieces that have come down to us of the *Inventio*. . . . It's kind of a very minor point, but she does at least suggest that the *Inventio* was true at least as far as this was concerned.

And the *Inventio* then talks about going west from this point and discovering several bays, one of which could well be Hudson Bay. Some of the cartographers that I was talking about have drawn Hudson Bay in their maps—in sixteenth century maps—and it would appear to be Hudson Bay. And a couple of cartographers refer to the *Inventio Fortunate* as being a source for the material for that northern area. So there's a reasonable connection between Hudson Bay—which was not known to exist or was not technically discovered until about 1612 by Henry Hudson—and these maps, which go at least as far back as 1507. . . . [So] we see this bay on maps as far as 1507, [and] we see people talking about their mapping of the area and connecting it with the *Inventio Fortunate*—therefore, there's a reasonable connection that the author of the *Inventio Fortunate* went into Hudson Bay, which puts him just upriver from Minnesota.

GONZO SCIENCE: It seems that the skeptics have a hard time accepting that anyone could have gotten that far that early. The very notion of Scandinavians exploring Minnesota in 1360 seems absolutely ludicrous to them.

ZALAR: Yeah, you go back to the very first scholar that examined the stone—Breda, back in 1899—and he puts down as one of the reasons that he just does not believe that the stone is real is the "utter absurdity of the story," something along those lines.

And yeah, there is that block, even with myself. I hear about this and I go, I wonder what the reasons are for thinking it could actually be authentic. The story seems absurd, to think anyone could have actually got here at that time—until you start doing the research.

And I think a number of these skeptics are specialists in a particular area, so they have a limited field of vision. Most of them are linguists, and they don't expand their investigation to include the historical aspects, or the geochemical aspects, or any of the other sciences.

GONZO SCIENCE: It might take more of a generalist to put all the pieces together.

ZALAR: That's one thing I've been trying to do in my discussions with Knirk or [William] Fitzhugh [of the Smithsonian Institution], to show them that there are a lot more pieces than what they've been looking at. Okay, maybe the pieces of the puzzle up in this area don't seem to go together, but all around that area, they do seem to go together. You have to explain these other pieces. You have to explain how someone could have gone into the area prior to the settlement of the region, because that's what the science is saying—physically, the stone had to be in the ground for [at least] 50 years in order for the kind of weathering that has occurred on it to have occurred [which exonerates Ohman and any of the area's settlers]. And you know, [the skeptics] say, "Well, that doesn't matter, we can ignore it because we've got this knowledge here that says it's impossible."

GONZO SCIENCE: They're saying linguistics trumps geochemistry.

ZALAR: What I keep pointing out is, even if they're right, and this is impossible or seemingly impossible, you've got another set of evidence over here that says it's seemingly impossible for it to have been forged. And what we have are these two masses of evidence, and what I would really like to see would be for people to come together under the Minnesota Historical Society or the

Smithsonian—some agency—to try to get a number of objective people to sit and simply review the evidence, and to hear the pros and the cons as put forth by the experts, and to make judgments regarding the evidence, the legitimacy of the evidence, and possibly come to some conclusion.

But we need a standard set of what we know about the stone. Because people on the one side are putting forward all their evidence with whatever bias they have, and we have people on the other side putting forth their evidence with whatever bias they have, and you hear claims and counter-claims being put forth, and you need someone to sit down and adjudicate that. That was suggested as early as 1910 or 1911. *The Journal of American History* suggested that the Smithsonian form a group to investigate the evidence…it was suggested at that time that a panel be formed to look into this and see what could be seen. And that has never been done by the Smithsonian; the Minnesota Historical Society put out their report in 1910 which was pro-Kensington Runestone. But as far as simple unbiased groups: like I said, just hearing about it, you go, "Well, that can't be right." That's the first thing that pops into just about anyone's mind…

GONZO SCIENCE: What's your take on the "AVM stone" hoax admission?

ZALAR: My opinion on the AVM stone was to get it into the lab and see what's there. The only interesting things that I had seen in the AVM stone were the three runes under the AVM which were not the same as the runes on the Kensington stone. They were cut differently, the letters were different, one of the letters was not even used in the Kensington stone at all, another one's backwards, another one has got a shorter staff and was kind of different. All I knew about it was that whoever cut this AVM stone was not the same person that cut the Kensington Runestone. That's all I could say going in.

Now, it is disturbing that [the hoaxers] did not come forward immediately after the announcement of the finding of the AVM stone. I think that was highly unprofessional, especially for someone who was a professor of Germanics, to allow this to go on for nearly three months after it was first reported found. And apparently she knew about it fairly quickly. It cost me time, and not so much money, but I am going to have to redo my booklet a little bit and put an insert into the booklets that I already have completed, discussing this.

So from that standpoint, it's quite aggravating. As far as whether it has any great influence on the Kensington Runestone—I don't see it. On the one hand it did bring the Kensington Runestone back into the public eye for a while, which I guess was a good thing.

GONZO SCIENCE: Yeah, but then to attach the hoax idea to it again…

ZALAR: And there was another hoax back in the 1940s, I think. Someone else carved a runestone and buried it in his field, and dug it up again a little bit later. And again, people went, "Oh this is interesting, it looks like it might be proof of the Kensington Runestone," but after a while, people started going, "Well, it really doesn't look much like anything," and [the man responsible] finally admitted that he had done it as a hoax, as a means of trying to disprove the Kensington Runestone. As if doing this really proves anything. It was the same thing with the AVM stone, except that it was a longer period of time before anyone found it, so it had been able to weather a bit more.

It was actually discovered in I think 1995, by a group that was out there, looking for any additional evidence for the Kensington Runestone they could hopefully dig up. It was found, it was documented, [and] it was [dismissed] as a hoax at that time.

Apparently the intervening six years [since 1995] had weathered it a bit more, or lichens had grown in areas they had not, and that looked like evidence that it was older. And who knows, if people had found it a hundred years from now, certainly it would have been heavily weathered at that point, and by then all the hoaxers would have been dead.

GONZO SCIENCE: Wow, you're right—that could have been really bad.

ZALAR: Yeah, it could have been really weird.

GONZO SCIENCE: The Kensington stone seems to have survived this hoaxing attempt. It seems a little childish, like they're so sure that the Kensington stone is a hoax, they wanted to provide another hoax to show how easily it could have been done.

ZALAR: You have to have absolutely no respect, and anyone who's done any work on the Kensington Runestone can at least show reasons why people think that it's real, even if they disagree. Anyone that's seen the 1910 report, with Winchell and two other geologists pointing out that it must be [a minimum of] 50 years old, have got to either be ignoring this, putting up a mental block against it—that's possible.

GONZO SCIENCE: Or disparaging Winchell's credibility.

ZALAR: You can't disparage his credibility. The awards that he won, the fact that they've named a building down at the university after him—and not because he donated a lot of money. The geology building at the University of Minnesota is there because Winchell is the greatest geologist that has ever come out of Minnesota. His work stands for itself. He was the editor of a major scientific magazine, he was the state geologist, he won awards at international conferences, he was president I think at one international conference. You don't get that by being stupid and prone to looking at pranks and thinking they're real.

And I've heard that argued briefly in one of my Internet discussions—"Oh, this Winchell must have been a flake that was doing that sort of thing for a living"—until I pointed out that, no, this is what was out there, it was confirmed at that time by the Wisconsin state geologist, Hodgkiss. As a matter of fact, his limit of 50 years is what I'm using. Winchell thought it was likely to be 500 years old, from his understanding, and his examination.

GONZO SCIENCE: So they were saying, "50 years at the very, very least."

ZALAR: At least. For [Warren] Upham and other geologists . . . [Upham] was examining it at the same time as Winchell. They worked together on the 1910 report. He thought it was probably hundreds of years old from the weathering. Hodgkiss said it was hundreds of years old, at least 50. So when I say 50 years old, I'm using the absolute minimum that any of the geologists have come up with.

One thing that bothers me is when someone says, "All the experts agree that the stone a hoax." I'm going, "Well, all the experts agree that the stone must have laid in the ground for at least 50 years." That's the only place where we find the experts agreeing. We find people like William Thalbitzer, S.N. Hagen, to a certain extent Holand, Dr. Nielsen, Dr. Hall, all saying that, "Look, this stuff could well be from the fourteenth century." So there is no "all the experts agree" on that point. And these are the people that have written papers, so we know they've at least examined it, not just read someone's article that quoted someone else who'd examined part of what someone was talking about and came up with this opinion and have that reflected up—that's not an expert opinion. You can't give an expert opinion until you've done the research. And if your "expert opinion" is that all the experts are agreed that it's a hoax, then you haven't done your research, and so I guess you can't be considered an expert.

GONZO SCIENCE: Maybe they mean all the linguistic experts? But then, that's not even true.

ZALAR: The Smithsonian did do a slight change in [their] book after I presented my list of errors. They did a second printing, and there were a couple of places where they did some—not in some of the major stuff; that would have required a whole rewrite I guess, and that wasn't something that I guess they felt they needed to do. But it seems to me that, and I'm really not sure of this, [Brigitta] Wallace [who wrote the section of *Vikings: The North Atlantic Saga* for the Smithsonian that Zalar had to correct] had had to back off of saying "all the linguistic experts" to something like "all the Scandinavian runologists" or something like that; having to box it down, because [Scandinavian linguist William C.] Thalbitzer was a trained linguist, and he thought the stone was real. So you've got to eliminate "Scandinavian linguists" from that group; you have to box it down to "Scandinavian runologists." So okay, all the Scandinavian runologists who've written about the Runestone think it's false—not all the linguists mind you, and not all the Scandinavian linguists—just this little box that kind of sounds good, but really doesn't mean all that much.

It was a runologist that was reviewing Nielsen's paper [for *Scandinavian Studies*]…if this runologist now believes that the Kensington Runestone is real, or at least that the evidence does not prove it to be a hoax, that it's still questionable—then there's another chink off of that as well.

There was a letter written by a fellow—he wasn't a linguist or anything like that, I think he was an historian—Frederick Brown, back in 1910, to a member of the [Minnesota Historical Society's] Runestone Committee. And he pointed out that if there is any doubt whether or not it's linguistically correct, then you've got to start going to the physical evidence and examining that. He said unless it's decidedly clear that the language is impossible, then you start looking at other factors.

And I think he said something to the effect that you've got to watch out because laws are not only validated by new findings, but new findings change the laws. He didn't give any opinion at any point one way or the other about whether the Runestone is real or a hoax. He just wrote this to someone who was examining the Runestone, the point being if there's any question in the linguistics, you have to check the other factors.

I think that there has been—from Dr. Fossum back in 1911 who wrote a newspaper article going word for word through the Runestone, right up to this current debate in *Scandinavian Studies*— there have always been people that are questioning whether the linguistics are wrong, and they have published papers on that, and they have said that the linguistics are reasonable for the fourteenth century for a party like this. The linguistics, and the runes, and so on. So you can't trust [linguistics] as the be-all, end-all. And when you look at the people that have come out against it, they are virtually all linguists.

Seaver might be an historian that has come out against it, but I don't know of many historians that have come out against it. I have known a few that have said I think the stone is real, or it seems reasonable: Thor Heyerdal, Farley Mowat, several others, those are the memorable names. And others have come out in support of the Runestone from their analysis of the information.

GONZO SCIENCE: From a historical perspective, it's really not that great of a leap. Here you have this tough, sailing culture, hard as nails, exploring all the time…

ZALAR: It's coming out now that they were interacting a lot more with the natives; they were certainly going to the North American continent to cut wood, because they didn't have any wood up in Greenland.

GONZO SCIENCE: These are the folks from the fourteenth century western settlement of Greenland?

ZALAR: Yes, the Greenland colony; they were going out and doing that. Bits and pieces have been put together; they were up north of Baffin Island. So they were working this area, rather than being a bunch of Norse who were all stay-at-home types, wouldn't venture out from their colonies, or rarely went between colonies or anything like that—which is kind of how they're looked at, I think. And now we do know that they were venturing forth, interacting with the native population in the area. I don't know if there was a lot of trade going on, but there are indications of trade. They were going up and down the coast of Greenland, to many islands, and even over to North America. They were going to all these places, and that's a bit more than we had thought before.

Why couldn't an expedition [get here], especially if it was fitted out like King Magnus was saying: "Take the royal ship and the men that you think are best, even from my own bodyguard." Now, we don't have any further knowledge of that outside of these orders that Magnus gave, and Magnus was overthrown the next year, so we don't even know if that trip started.

Though there might have been some church involvement regarding that. There are indications that Sweden was behind on her tithes to the pope, and this may have been some way of getting the pope off their backs, where you're out making sure that there's Christianity out here, and maybe trying to spread the word, or finding out whatever and sending that information back to the pope.

So it may have been that even if Magnus had been deposed, it still would've had to have gone on. I don't know the history of that period as in depth as I would like to. There aren't a lot of books out there about Sweden in the 1360s.

GONZO SCIENCE: Maybe you should write the first one.

ZALAR: It's a time period where Sweden was large. Magnus was both king of Sweden and Norway, and had a lot of Finland, and they bought some territory on the Swedish peninsula from the Danes, where the Danes had used it. So he had a large chunk of territory. He was crusading against the Russians, maybe open to expanding his influence there.

So it's certainly not impossible that if he was this expansionistic, that he would have looked to Greenland, especially as a— I mean somewhere up there, there's Asia, Cathay—and Greenland may have been real close to it. Greenland had white falcons, and according to one report, the Mongol king, the Khan, had white falcons. Maybe Magnus heard this and kind of put two and two together, maybe he had descriptions of the Eskimos, which look like Asiatics. It might have been like, "We might be pretty close here. Let's see if we can find Cathay, find China."

That's my thought of why the expedition would have gone as far inland as they did—that there were reasons at that time for someone in his position to think that he might be able to get into some very lucrative trade via Greenland.

Of course it's hard to look into his mind. It could be like he said, "I want to preserve Christianity out there, and that's it. If you can poke around a little bit more, or look for these lost colonists, or whatever, fine, go ahead—but don't make too big a trip of it." And they just took off on their own and went, "Hey, this is kind of neat—

let's see what's around the next corner."

You can go up the Red River as far as Fergus Falls without any real difficulty before you hit any kind of really major rapids or anything like that. And Fergus Falls is, what, 40 miles from Kensington, roughly? I've got no problem with the idea that the massacre may have occurred there, at the "day's travel." Walking south along the tree line, out of sight from Indians at that point, because the tree line does go right—roughly—from the Fergus Falls area down to the Kensington area as far as I've been able to determine. So that's a possibility.

GONZO SCIENCE: What do you make of *The Kensington Rune Stone: Its Place In History* by Thomas E. Reiersgord? Reiersgord says that the stone was carried around by Indian tribes, and they got it to the KRS hill, where it was eventually found by Olof Ohman.

ZALAR: It makes for an interesting possibility but I don't think there's a lot of fact that goes along with that. For instance, he says that the plague wiped out these ten people. And the idea of them getting this far in, having carried the plague for two or three years before manifesting itself, and only manifesting itself in these people at this time, and all dying at once…that seems pretty shaky.

He thinks they came in by Lake Superior and then a number of backwater routes—you know, you can't get up the St. Louis; around Jay Cook Park there are massive falls and stuff like that. I admit he did present a route where they might have been able to follow, if they had Indian guides to navigate them from here to here, to know where the portages were from there to there, and then go upriver here, and go off the river here. But it's too complex.

I've always gone for the simplest explanation. The simplest is coming in via the Red River—the Hudson Bay to the Red River. That's pretty direct. The conclusions of Reiersgord's book are not that wild, but they're just a little bit too far for me to really accept, unless there's more tangible proof. Like the Indians wrapping the stone in cloth of some sort and carrying it around, and only showing it on special occasions, like it was some religious icon of the orthodox faith or something like that. You know, if there was evidence out there that they did this a lot with religious artifacts, that they'd cover them and carry them around—but as far as I can tell, there's just no evidence that supports the idea.

Recommended Reading: Michael Zalar's "Factual Errors in Chapter 29, Vikings: The North Atlantic Saga, Regarding the Kensington Runestone" *NEARA Journal 34:1 (Summer 2000) p 7-10.* Richard Nielsen's paper, "Response to Dr. James Knirk's Essay on the Kensington Runestone," is the first comprehensive treatment of the language of the stone to be peer-reviewed. It may be found, along with many supplemental papers, in Barry Hanson's *Kensington Runestone: A Defense of Olof Ohman, the Accused Forger*. Hanson's non-profit archaeology organization's website is Archaeologyitm.com, where one may find his book.

CHAPTER 24

Interview with Velikovskian Researcher Charles Ginenthal, Part 2

VELIKOVSKIANISM IS SO INTERESTING TO ANOMALISTS because it spills over into other disciplines besides astronomy. After all, Velikovsky began his critique of science as a psychiatrist-historian and then built an alternative worldview by following a trail of anomalies through astronomy, cosmology, geology, and archaeology.

Velikovsky discovered that the conventional chronology of history clashed with his own reading of events, a reading based on the idea of recurrent cosmic disasters befalling humanity. The more he looked into it, the more it began to seem as if the traditional history—in particular, that of the Near East—was full of holes. Specifically, the conventionally accepted historical timeline appeared to be artificially inflated with years.

Charles Ginenthal came to this conclusion as well. As he relates in his article "Science, History, Ramses II and Velikovsky,"

> I had assumed that the history was sufficiently well documented, so that if one were to revise the established chronology, the historical documents, the "details" would support it...[But] the documents are so limited in their lack of fullness, merely "rags and tatters," of that historical time that they cannot be held as the ultimate arbiters of that chronology...Velikovsky introduced the concept that scientific evidence overrules histori-

cal documentation. This…is the process this author [Ginenthal] accepts as the proper methodology by which chronology should be evaluated and ultimately determined (p. 60-64).

In this segment of our interview with Ginenthal, he details some of the archaeological anomalies and controversies of Velikovskianism.

GINENTHAL: There are very important discoveries being made in ancient history by Velikovskian researchers who have gone beyond Velikovsky. One of them is Professor Lynn E. Rose.

Now the foundation of Egyptian history is based on astronomy. That is, a particular star rises at a particular time in the life of a particular king—we can retrocalculate back to that time, and we can say that this king lived at a particular time in terms of astronomy. There's no question about it.

They call that Sothic dating, that is, a particular star rose—they call it a helical rise—it comes up over the horizon and the sun comes up immediately after and blots out its light. And 1,460 years later, because the Egyptians lost a quarter of a day every year, their year was only 365 days, thus this star would rise again—not precisely, but around the first day of the first month. And this star would rise again around 1,460 years later and they would get another helical rising, which they called the "great year." And that's how they dated Sestrosis III. He's of the twelfth dynasty. However, they also had lunar holidays or lunar celebrations that were dated to him and to other kings, the pharaohs of the twelfth dynasty.

And the question was, can you make these lunar dates coincide with the astronomical date of the rising of Sothis? And therefore you would have it all nailed down.

A lot of people were concerned about that. Archie Roy was one of them. And a man named Edgerton said, "We have to make this stuff fit. We don't really have a solid date until we fit the lunar data with the rising of this star." Some of the world's leading archeoastronomers—Parker among them, he's considered one of the greatest—tried it, as did Luft, Krauss, and Bauchart. They all tried to make the data fit and they all claimed that it fit pretty well.

The problem was, Professor Rose looked at all their work and he published an article in the *Journal of Near Eastern Studies* put out by the University of Chicago. He showed that they didn't have fits; since the moon repeats itself, if you get one date, down the road you're going to get other dates. And therefore what they were getting was 42 percent fits, 38 percent fits, 40 percent fits, 51 percent fits—they were not getting 85 percent, 90 percent, 95 percent fits. In astronomy, you would expect to get at least 90 percent. In other words, Egypt may have a cloudy day here and there, a rainy day, or a dust storm, so you wouldn't get all the fits but you'd get a pretty good number of fits.

What Rose did is say, "This doesn't look like it goes that far back in time." Thus he moved Sestrosis III and the date of the helical rising of the star Sirius—which they call Sothis—he moved it forward in time by 1,477 years, and the lunar data fit, everything fit, and he published this work.

Now, this is an astonishing piece of work. That is, the arguments against Velikovsky were always based on astronomical retrocalculations, that is, on calculating orbits. And these critics said, "These orbits couldn't be, and the history doesn't support it."

Now here is retrocalculation, that is, Rose used their work, their corrections, their approach, he used all their data, all the corrections of the data, and their methodology. And he's the only one who has a fit. That is, of the 36 he got 34, and one of them he could explain in terms of the scribe writing "up to" instead of "and beyond," and the other one he could take as a point of poor seeing. That meant he got about 94 percent, which is what you would expect in astronomy—good astronomical retrocalculations. This of course is upsetting to the astronomical community.

I keep finding scientific evidence that contradicts the paradigms of history is totally ignored and thrown out the window. Then people who attack Velikovsky say, "There's no evidence that the history is shorter." And yet here's a man who's showing you that the history is distorted.

Another follower of Velikovsky, who also doesn't go along

with the history of Velikovsky, but is more extreme than Velikovsky, is Professor Gunnar Heinsohn at the University of Bremen in Germany.

His argument is that history is much shorter than Velikovsky's. Velikovsky wanted to shorten history by about five or six hundred years. He wanted both the beginning of history and the end to be shorter...These people like Rose and Heinsohn are bringing it down greatly.

Heinsohn said, "Look. The stratigraphical research—that is, what you find in the ground in Mesopotamia—does not support a long chronology, a long period of chronology." A group of people settles in a place, then go away, and according to historians about 700 to 800 years or 1000 years later, another group of people come and settle in the exact same place. And what they have there is what is called a settlement gap. Now, Heinsohn said, "Look at the excavation reports—there are no settlement gaps! You're creating history where there is none. There are no settlement gaps between certain civilizations."

He says, "You're actually creating civilizations!" He says, "Look. No one in ancient times ever heard of the Sumerians." He said when they dug them up, all of a sudden, every bit of material for the ancient Chaldeans simply disappeared. He said, "You can't find anything for Chaldeans." He says, "Can't you understand that you dug up the Chaldeans and not the Sumerians?" He said the same thing about other groups: "When you dig these up the other people just disappear!"

Anyway, he said there was a group of people called the Mitanni. And another civilization called the Old Akkadians. He said the Old Akkadians settled in an area first, and 800 years later the Mitanni settled on top of them. And you have places where you find this, you know, these kinds of things. And he said, "Look, there is no settlement gap between the Mitanni and the Old Akkadians. Because the Mitanni are the Medes, and the Old Akkadians," I believe he said, "are the Assyrians." The Medes and the Assyrians follow one after the other.

Anyway, he was going to meetings, to historical meetings,

and getting everyone upset with what he was saying, because he's a professor of economics but he was going to historical meetings. And at one of the meetings, Edith Perada from Columbia University almost spit on him. She was frothing at the mouth. She's the great Assyriologist in this country.

But Heinsohn was corresponding with one of the leading Assyriologists in Germany and asking him these questions. And the fellow decided to come to a meeting. He's a very, very old man. He's like the Nestor of Assyriology in Germany. He told Heinsohn, "Take my arm, I want to talk to everybody." And when he got up to talk, people listened. And he said to the people, "Look, Heinsohn has been raising issues and you've not been treating him very fairly or very nicely. And I've been dealing with him for years and he's got a lot of good questions. Now, if Heinsohn turns out to be right, your students will curse you."

Of course, it was incredible to hear that from him. Perada then came over to Heinsohn and says, "When you come to New York, we'll discuss your insanity." But it was out there. The whole thing, his theory, was out there. The German people have been reading Heinsohn's work; he's a well-respected writer and thinker in Germany.

Anyway, a fellow by the name of Wilfried Papa decided on his own, he said, "Look, we're going to test it." And so he was part of a team that was going to test the settlement gap between the Mitanni at the top and the Old Akkadians at the bottom, at a site to see if there was a 700- to 800-year settlement gap between them, in 1988. He said, in 1988 after the first dig, "I don't understand, there's no settlement gap."

This upset everybody. They came back in 1989, and they put three holes in the ground; they call them soundings. And they looked everywhere for this settlement gap. They brought along a geologist by the name of Ulrike Rosner to look for windblown layers. Because if the place was left alone for 700 to 800 years, wind would blow dust into all the crevices. These are called aeolic layers. And what they found after all three digs is that there was no settlement gap.

They don't want to even talk about it. I have a letter from

Wilfried Papa saying, "I went there to prove you wrong, and I can't, the evidence shows you're right. The contract seals at the bottom in the Old Akkadian are the same as you have in the Mitanni right above. They're exactly the same. The foundations from one age go right into the other. There is too much evidence and there are no aeolic layers."

GONZO SCIENCE: So what does this mean?

GINENTHAL: Astronomers were able to say, we can look back, way, way back in time and prove that the solar system was as it is, because the ancient Egyptians and the ancient Mesopotamians could point to Venus being in the sky. But if you bring these civilizations down closer to the present, the whole argument collapses. Their arguments fail. And that's what's essentially happened.

GONZO SCIENCE: So the astronomers and the Egyptologists are leaning on each other, but it's a house of cards?

GINENTHAL: I believe so, yes. Many people want to get rid of this evidence, now that Rose has shown that Sothic dating is wrong. Some people are saying, "We don't even have to use Sothic dating, it's not important. We have all these other forms of dating."

GONZO SCIENCE: Of course they do.

GINENTHAL: But all their other forms of dating have enormous problems...They have similar problems in other areas. I'm sure you're familiar with the problem of cutting hard stones without iron in the Copper Age.

GONZO SCIENCE: Why don't you explain it.

GINENTHAL: In the pyramids there are stones that are made of granite—very hard, rose granite—and there are statues from the Old Kingdom which are supposed to be from 2500 years B.C., which were carved at the time when they only had copper. They didn't even have bronze at the time.

And Heinsohn says, "Look, you can't carve some of these statues. One of them is diorite, which is an extraordinarily hard stone to cut with copper."

Flinders Petri looked at this stuff and said, "Look, we have saw marks right here in the granite—you can actually see the saw mark. They were cutting with saws with teeth. And look here where the drill marks are, you can actually see the drill moving down into the stone. You can't do that with copper. You have to have something like steel, or diamond, or topaz, and none of this existed in Egypt in 2500 B.C."

No steel, no topaz—they've tried every single thing. They came up with a theory that you could do this by taking copper and sand, and as the copper moves back and forth on the stone, the sand, which is very hard, will cut the stone. But, it will not leave teeth marks, that's the problem—it'll leave a nice smooth edge, it'll be almost as smooth an edge as you'll get on polished granite, but not as smooth—you'll get a very smooth edge but you won't get saw marks. If you took even a round stone, or a round piece of copper, and you put it on a piece of granite or this other material, and you turned it and turned it and turned it and pressed on it, and kept wetting it and turning it, eventually the sand will drill into it, but you won't get the drill marks around the edges. So they don't have any way to do this.

But according to Heinsohn, if you move the civilization of Egypt forward, we have steel and then they can cut it. If they have diamonds, they can cut it. If they have topaz, they can cut it. But you have to rewrite the history in order for this to fit.

There's another fascinating problem I came across, which deals with the ancient Sumerians. I'm preparing a book on this called *The Pillars of the Past* and I'm giving you some of the information from it. They supposedly lived in southern Iraq for almost 3,000 years. They had an early period called the Ubaid period, which goes from 4500 to 3500 B.C. And then they had a period called the Uruk, which goes from 3500 to about 2900 B.C. And then you have the Pre-Dynastic period, then the first dynasty, second dynasty, third dynasty. And then you have the Akkadians taking over but controlling the same area.

So [people] lived there for about 3,000 years in the conventional chronology. They carried on irrigation farming, but the problem with irrigation farming in that part of the world is that the water table is very close to the surface. If you have a high water table and hot long summers you run into a problem.

The water doesn't have to come completely to the surface. If it gets within 18 to 20 inches of the surface, capillary action through the sand grain brings the water closer to the surface. The hot surface heats the water, and the water evaporates and flows upward into the atmosphere.

But you can't get rid of the salts in the water. The salts go onto the land, and that's one of the major problems with irrigation farming in that part of the world today. All throughout Iraq, and Pakistan and these countries, they're really in serious trouble. Unless they figure out something to stop this from happening, their whole economies will go crashing to the ground.

Anyway, somebody asked how long can you carry on irrigation farming. And the answer, according to one expert who at their very best, examined the northern part of the plain—that is, the place where the water table is not so high—he said, using the fallowing method, that is one year you plant, the next you let weeds grow and they suck up the water and bring the water table down—and you go through that whole business—and he said, "The land will become useless in this part of Iraq in 450 to 500 years, but on the plains it's [even] shorter." So you can have agriculture for no more than about 300 to 400 years and maybe less. And yet we're expected to believe irrigation agriculture was being carried on for about 3,000 years and intensively for about 2,000 years. It doesn't make any sense at all.

GONZO SCIENCE: So it's much more like what Velikovsky said it was.

GINENTHAL: History is shorter. One of the aspects of Velikovsky that is fascinating is that he uses history.

Ramses III, I think—they discovered a palace that he had. And it had tiles in it. And Ramses III is dated to the second millennium B.C. But they found a group of tiles, and the tiles had Greek letters on them. Not imitation Greek letters, but straightforward Greek letters. And they've been going crazy trying to figure out where they could get these. They tried different forms of hieroglyphics; they stood on their heads. It doesn't work. If Ramses III, as Velikovsky suggests, lived in the first millennium, about 400 to 500 B.C., then these letters were available. The Greeks were making tiles and selling them to the Egyptians.

You know, things like this that are so clear cut, it's absolutely fascinating...It's all like this! Not that the anomalies—and there are expected to be anomalies in any theory—but when the anomalies all fit together, when they all point in the same direction, when Sothic dating says first millennium, and the lack of settlement gaps in Mesopotamia brings the history down to the first millennium, when iron says you can't cut this before the first millennium, and there are other problems with tin, and other kinds of things, which they just can't explain in terms of their chronology. And these are all scientific and technological.

But when they find scientific and technological evidence that supports their theory, it's hailed—it's immediately accepted. If they find scientific evidence—solidly supported scientific evidence—which contradicts their theory, then they go off into all kinds of tangents, and they begin to invent ad hoc hypotheses...

The reason I support Velikovsky is that when I look at the evidence—and this is in terms of several pieces of evidence, all corroborating a concept, not just one piece points to it but say seven, eight, or nine pieces of evidence point to the same thing—then it seems to me that the corroboration is a powerful form of support.

And what we talked about before, say Mars—I mean, why would it have an ocean at its present distance? Why would it have river valleys if it was always in its present condition?

And you take a look at the erosion on the planet—no one talks about the erosion—you can read about this in my book *Carl Sagan and Immanuel Velikovsky*—they can quantify the erosion. They know the erosion is there, they've tested it. The erosion says Mars should be covered by sand dunes, and nothing but sand dunes. And yet they see rocks, and the rocks are hardly ventifacted—that means they're hardly worn by windblown sand. And then they see rocks which begin to show the beginnings of ventifaction of windblown sand.

When they look at some of the gasses on Mars, like N15, this doesn't make any sense in terms of the present history in terms of Mars.

All of these concepts, all of these forms of evidence, corroborate each other, they all fit together with Velikovsky's theory. And that's what keeps me going.

I would like very much first of all to have that theory tested in space, but I would like somebody with a magic bullet, or with a wooden stake, to stop it, to stop the evidence, and prove once and for all that every time we look at the evidence that's being presented—or I look into it, or other people look into it—it simply doesn't hold up.

Now, they can say that we're creating the ad hoc hypotheses, but if you take a look at all the ad hoc hypotheses that they are using to uphold their theories, then we're not doing anything of the kind.

Nobody has ever been able to show that George Talbot's temperature analysis of Venus is totally incorrect. What they do is they begin to poke fun at it, and to attack it and misrepresent it. Lynn Rose's work on Sothic dating has already been lied about by a major person involved in the field. That is, it was so upsetting, and he couldn't argue with the mathematics, so what do you think he said? "You're using too much mathematics!" And then he said that Rose equates Sothis—[what we call] the star Sirius—with Venus. And he said it four or five times, this fellow. And yet at the same time I went to the book and looked, and in four or five places, Lynn Rose says, "I don't believe that Sothis is Venus, I believe it's the star Sirius." And he says it in four or five places. So why would a man like Anthony Spallinger, who's familiar with the field, go out of his way to say such a thing? That's not science. That's dirty politics.

Recommended Reading: *Worlds in Collision* by Immanuel Velikovsky; *A Guide to Velikovsky's Sources* by Bob Forest; *Scientists Confront Velikovsky* ed. by Donald Goldsmith; *Scientists Confront Scientists Who Confront Velikovsky* ed. by Lewis M. Greenberg and Warner B. Sizemore; *Carl Sagan and Immanuel Velikovsky* by Charles Ginenthal; *Science Frontiers* by William Corliss; "Science, History, Ramses II and Velikovsky" in *Velikovskian*, Vol. 5, No. 4, by Charles Ginenthal.

PART 7:
Gonzo Energy and Environment

CHAPTER 25

Four Recent Signs of Impending Global Catastrophe

BY WAY OF INTRODUCTION, WE HERE PRESENT FOUR signs of impending global catastrophe, a thin sliver of all available signs of impending global catastrophe. Just pick up the paper.

We would argue that it is time to take personal responsibility for our planet and start doing whatever it takes to ensure that something is left in 50 years — and that what's left is worth having. We would also argue that the state of the environment is so bad that it is too late for merely being cautious, a la the so-called "Precautionary Principle" so in vogue now among the environmentalists. It's more like time for an "Actively Panicking and Scrabbling to Keep From Going Over the Edge Into Apocalypse Principle."

This chapter leads through some pretty bleak landscapes, and then emerges into a more hopeful place. Enjoy.

1. The North Atlantic Ocean Will Be Fished Dry in Eight Years

Fisheries scientists at the University of British Columbia completed the first comprehensive survey of the entire North Atlantic Ocean, and they presented their findings at the 2002 annual meeting of the American Association for the Advancement of Science (AAAS).

Their results? The entire fishery could collapse as soon as 2010. It seems that since 1950, the fishing industry has tripled its efforts to catch fish, and yet their catches have sunk to half the 1950 levels. This is known by the scientific term "overfishing."

Could market forces help? Fish prices have risen to six times what they were in 1950, as measured in real terms. You might think that with high prices, people would just stop buying fish, and the fishers would go out of business, hopefully before every last fish is removed from the ocean (which will also put the fishers out of business, obviously).

In fact, fish prices should be a good deal higher than six times the 1950 price, except that the U.S. government, in its wisdom, slathers $2.5 billion a year in subsidies over this ecologically devastating industry. It's like, hey, what we need is more fishing up there!

There have been limited efforts to protect certain fisheries, which only make the fishing fleet move their overfishing operations elsewhere. In an interview published in *New Scientist*, Andrew Rosenberg, a fisheries scientist of the University of New Hampshire, likens this to "shuffling the deck chairs on the Titanic." According to Rosenberg, nothing is being done that can stop the decline in time. Apparently the UN has some voluntary initiatives going, and some other things are being discussed, but those things could take longer to get started than the few years we have left. According to these fisheries scientists, the only thing that's going to work is to reduce the size of the fishing fleet and to make large swaths of the ocean into non-fishing areas.

But hey, the Japanese have hit upon a solution. They are now saying that *whales* are the real problem. According to the whale-hungry Japanese government, those pesky whales are the ones doing the overfishing! So the obvious solution to declining fish populations is to kill more whales. Brilliant.

2. The Amazon Rainforest Is More Damaged than Previously Suspected We all know about the slash-and-burn routine that has destroyed 14 percent of the Amazon rainforest. There's more to it. A study presented at the 2002 AAAS annual meeting has shown that not only is 14 percent of the rainforest burned to the ground, but 50 percent of the remaining rainforest is being heavily damaged by the smoke.

This adds a whole new dimension of disaster to the problem. It seems that the smoke is absorbing almost half the sunlight that would have reached the forest otherwise, which screws up photosynthesis in a big way. Not only that, but the burning produces ozone which, darn the luck, doesn't obligingly float right up and conveniently plug the ozone hole over Antarctica. Instead, this ozone (which is toxic to plants) is blowing all around (up to 1500 kilometers away from the burning) and reaching levels seven times higher than normal, poisoning the forest the whole time.

The Brazilian government is tackling this problem by trying to get farmers to—get this—stop burning down the rainforest.

3. Sea Levels Are Rising Four Times Faster than Previously Believed That AAAS 2002 annual meeting announced even more signs of the Apocalypse. Two scientists from the University of Colorado at Boulder claimed that past estimates of sea level rise have been drastically underestimated. Turns out past calculations have figured between one and 23 centimeters of rise by 2100 (with roughly one meter of shoreline gone for each centimeter of sea level rise). These guys are saying that, hello, you forgot to count all these melting glaciers in Canada and Alaska, and that really, we're looking at a rise of more like—gulp—89 centimeters. Time to sell off your beachfront properties folks, and maybe buy a houseboat.

4. "Coral reefs older than the Pyramids are being smashed to bits"—New Scientist, February 26, 2002
European fish stocks are collapsing (sound familiar?), so the trawlers are heading out to new territory—4,500-year-old coral reefs, home to the great majority of ocean species. The trawlers drag miles-long nets, held open with weights that weigh up to a ton. Dragging across the sea floor, the weights turn everything to shit, including whole ecologies we know nothing about, and never will. But hey, what's for dinner?

CHAPTER 26

Homer Simpson Is Real: World's Dumbest Nuclear Accidents

NUCLEAR ACCIDENTS ARE DUMB BECAUSE NUCLEAR power itself is dumb. From where we're sitting, if you're going to handle the most dangerous material in the world, you must be an idiot if you're going to handle it in any way except for the safest possible way. And don't tell us we're being as safe as possible with our nuke tech, even since September 11. Like the way all those Iraqi nuclear sites went unsecured after Operation Iraqi Freedom? What was that all about? And the way Greenpeace had to shore up the containment efforts with barrel exchanges for people who were drinking out of looted uranium barrels?

You've simply got to treat nuclear material like it's Kryptonite and we're all Superpeople. Can you imagine that the High Council of Krypton Elders would have stood for the situation that we have here on Earth? There's Kryptonite everywhere—leaking out of barrels, draining into water, venting into the air, and pumping deep into the planet—no wonder Krypton exploded.

It is openly recognized in the business world that the nuclear power industry is the worst managerial disaster of all time. And the military is right in alongside them with its nuclear disasters and plutonium in your lungs.

Correct us if we're wrong here, but this stuff is lethal by the billionth of a gram (we're not wrong), and they think nothing of putting it in perfectly ordinary airplanes that go around crashing all the time. You've got to have some kind of Superplane that never crashes. Isn't that only fair to ask?

At least give us some kind of Supersafe reactor without all

the human error, crackbrained experiments run amok, broken seals, burst pipes, and meltdowns. Those poor people. . .

Those people could be us. Who here knew we almost lost Detroit on October 5, 1966? Yeah people, it could happen. But would that catastrophe finally signal the end of nuclear power, which cannot be cleaned up, only dug up and reburied somewhere else? Find out who your president is and write him. Tell him, "I am against getting killed with nuclear stuff and I vote."

Now here for your amusement: The 14 Dumbest Nuclear Accidents of All Time (from 1945 up to 1979—you've probably heard about the ones after that).

1. Alamogordo Bombing Range, New Mexico, July 16, 1945. General Leslie Groves ignores the meteorologists. For political reasons, Groves rushes the first-ever test of the atomic bomb. It was to give Truman the upper hand at a conference of wartime leaders. The test accidentally "contaminated the air over an area as large as Australia" (*New York Times*, May 23, 1946). Moral: Don't ignore the meteorologists.

2. Los Alamos, New Mexico, May 21, 1946. Louis Slotin removes the safety devices. While testing a bomb core assembly that had lethally irradiated a friend of his nine months earlier, Slotin "improvises" and holds two nickel-plated plutonium cores apart by hand. He goofs it and they touch. There is a blue glow and Slotin drops the assembly to the floor. Although short of a thermonuclear explosion, the blue glow proves lethal to Slotin nine days later, but not before his arms swell up like Popeye's. Moral: Never "improvise" with plutonium.

3. SL-1 Reactor, Idaho Falls, Idaho, November 29, 1955. A technician presses the wrong button. Instead of pressing the "immediate" shut-down button during a high-energy test, he presses the "slow" shut-down button. It takes him a few seconds to press the right button, but in that time half the fuel rods melt into a radioactive slagheap. Luckily, they did not reach critical mass. The reactor guys kept the accident a secret from their bosses at the Atomic Energy

Commission, who learned about it from a reporter's question. Moral: Don't press the wrong button.

4. Kirtland Air Force Base, New Mexico, May 22, 1957. An officer loses his balance in the back of a B-36 bomber at 1,700 feet and catches himself by grabbing the H-bomb release switch. The bomb falls right through the closed bomb-bay doors and lands in a cow pasture, killing a cow. Although there is no thermonuclear explosion, the conventional explosive component of the bomb detonates and scatters radioactive debris in a crater 12 feet deep and 25 feet wide. Moral: Never grab the H-bomb release switch to catch your balance.

5. SL-1 Reactor, Idaho Falls, Idaho, January 3, 1961. The control rods are reconnected by hand. Three reactor operators named Richard McKinley, John Byrnes, and Richard Legg go down to the basement to manually lift the control rods out of the reactor to reconnect them. This was seen as a time-saving, streamlined procedure. Lifting the rods any more than 16 inches is enough to make the reactor go critical, but because it is being done by hand, they just eyeball it. McKinley lifts a rod 20 inches instead of 16. All three men are killed in the resulting explosion of nuclear fuel and radioactive steam. McKinley himself is actually pinned to the ceiling by the control rod. Their heavily contaminated bodies must be buried in coffins surrounded by a foot of concrete and three feet of packed earth. The building takes 18 months to decontaminate. Moral: Get a job outside of the nuclear industry.

6. Observation Hill, Antarctica, 1961-1972. A leaky nuclear reactor is built in Antarctica. This one is almost too horrible to be true. The U.S. Navy installs a small nuclear reactor in pristine Antarctica. Predictably, it proceeds to leak radiation. When the reactor is decommissioned in 1972, it is shipped to the States for burial along with 101 large drums of radioactive earth. Over the next few years an additional 11,000 cubic meters of soil and rock have to be removed and shipped to the United States. Now it is officially declared clean. Go Navy! Moral: Don't build nuclear reactors in Antarctica.

7. Nanda Devi Mountain, Himalayas, Winter of 1965. The CIA loses an eight-pound plutonium battery at the source of the Ganges River. The CIA, in partnership with the Indian government, attempts to set up a monitoring station on the mountain in order to eavesdrop on a top-secret Chinese nuclear bomb test. A crack team of civilian climbers is assembled from America and India. They are to run their temporary encampment with the aid of the latest nuclear gizmo, the SNAP generator, which is essentially a cone-shaped plutonium battery that you can wear on your back. Whoever thought of this got a raise. The team starts their climb but must abandon their gear, including the SNAP, because of bad weather. Winter comes and goes and they climb up there again to rendezvous with their plutonium. But what do you know—an avalanche has carried it all away and buried it beneath thousands of tons of rock and snow. Several expeditions have failed to locate the SNAP. There is official denial that the SNAP poses a threat, on account of it being so safe and well designed and all. This really inspires confidence. They send another SNAP up there in 1966 and abandon it again, though this time it is successfully retrieved. Moral: Put some kind of indestructible homing beacon in your SNAPs.

8. Seventy miles east of the Japanese Ryukyu Islands, North Pacific Ocean, December 5, 1965. Faulty brakes send a nuke to the bottom of the ocean. An aircraft with a one-megaton atomic bomb and faulty brakes rolls off the airplane elevator of the aircraft carrier USS Ticonderoga and sinks like a stone—pilot and all—to a depth of three miles down. The Japanese government is misled about the incident—hey, screw 'em, right? The eventual clamor for the bomb's recovery has met with no success. Moral: Where are those indestructible homing beacons?

9. Palomares, Spain, January 17, 1966. The United States drops four H-bombs on Spain. An American B-52 bomber collides with its refueling tanker at 30,000 feet above this coastal village of 2,000 people. The bomber's four H-bombs are scattered around town, one of them landing in the Mediterranean Sea. There are no thermonuclear explosions. But, of the three bombs that hit the ground, the conventional high-explosive components of two of them detonate, which scatters the vaporized plutonium bomb-cores all over the village. The dumb part of this particular accident rests with the media coverage of this near-nuking of 2,000 people. The media focus is put on the recovery of the bomb that fell in the water, and when it is raised everyone shouts "Hooray" and goes home. Meanwhile, the villagers are seriously contaminated, and after the accident, they were not even treated very nicely, which you might think would not be too much to ask. They have to fight for their own medical records and are denied access to them until 1985. The United States has no criteria for dealing with overseas accidents, and the villagers are still flapping in the breeze. Moral: If you get contaminated with my plutonium don't come crying to me.

10. Fermi Reactor, Detroit, October 5, 1966. The Atomic Energy Commission ignores the critics who say the reactor is unsafe. They build it anyway, resulting in a major accident in October 1966. Legal action to stop the reactor's construction is fought all the way to the Supreme Court, which gives the legal go-ahead. Various industry shenanigans play out during this time, including an Atomic Energy Commission-suppressed safety report. This is subsequently uncovered and stimulates a further hail of public criticism. The public is ignored, the safety data is ignored, and following a string of assorted "minor" radioactive mishaps, the reactor almost melts down three years later. Know anyone in Detroit? You're lucky. Moral: Don't ignore the safety reports.

11. Unidentified U.S. reactor, March 1968. Nuclear plant workers use a basketball to close off a water pipe during repairs. That's right, a basketball. They wrap it with two inches of rubber tape to increase its diameter and then inflate it in the pipe, which proceeds to blow the ball out along with 14,000 gallons of radioactive water. Moral: Homer Simpson, eat your heart out.

12. Tureia Island, French Polynesia, June-July 1968. The 60 inhabitants of Tureia are not evacuated during three above-ground nuclear tests by the French. This one is not the best example of stupidity per se, but more of a complete callousness, cynicism, and racism. The two French meteorologists on the island were evacuated to a hospital for a week, decontaminated and tested. One can almost hear De Gaulle with his undies in a bunch: "*Sacre bleu!* There's white people on that godforsaken island! Get them outta there!" Moral: Never trust the French.

13. Unidentified U.S. reactor, April 1969. Nuclear plant workers connect the drinking water tap to a 3,000-gallon radioactive waste tank. The accident is discovered when radiation shows up in the reactor's bathroom sinks and in one of the drinking fountains. Moral: If you ever tour a nuke plant, bring your own water.

14. Murora atoll, French Polynesia, July 6 through July 25, 1979. A round of French nuclear testing is beset with error, miscalculation, and poor judgment. First, a radioactive concrete bunker explodes. It, and many like it, had been used to test the results of chemical high explosives on plutonium, whereupon they were normally abandoned and sealed shut. This time, though, the geniuses over at the French military decide to clean and reuse one. During cleaning, the bunker fills with acetone fumes and explodes, killing one and seriously injuring five. This explosion, while not a thermonuclear one, nevertheless scatters plutonium over the entire island. A couple of weeks later, a bomb becomes lodged 400 meters down an 800-meter shaft and is detonated anyway. At 120 kilotons, it is one of the most powerful nuclear explosions ever. Three hours later the atoll literally crumbles and loses one million cubic meters of its outer wall, which sloughs off into the ocean. This causes a tidal wave that spreads throughout the island chain. No one is killed, and it's okay anyway because the French version of the Atomic Energy Commission claims the tidal wave was strictly of natural origin. Vive la France.

CHAPTER 27

The Two Greatest Nuclear Conspiracies

CONSPIRACIES, COVER-UPS, AND OFFICIAL DENIAL GO hand in hand with nuclear technology. Accidents with nuke tech are so deadly and long lasting that, if nuke tech is your bread and butter, you would be a fool to admit the vast, global extent of nuclear accidents. Public opinion will turn against you, so your best option is to sweep it under the rug, to lie and to obfuscate, and to get real cozy with your regulatory agencies. Please note that even now, this pattern is repeating itself with the dangerous technology of genetically modified organisms and their snuggly kiss-kiss relationship with the (U.S.) government agencies charged with their oversight. But that's a chapter in itself. Ladies and gentlemen: the two greatest nuclear conspiracies.

The Murder of Karen Silkwood, November 13, 1974

Here is the quintessential nuclear conspiracy of all time, American style.

This nuclear plant whistleblower was deliberately contaminated on several occasions by unknown persons from within the orbit of the Cimarron plant, which was owned and operated by the Kerr-McGee Corporation.

Even the food in Silkwood's home refrigerator was found to be contaminated, which showed she had been the victim of a breaking and entering, and that her food had been deliberately poisoned by company thugs.

When these scare tactics failed to dissuade Silkwood from pursuing her safety agenda, she was fatally run off the road on November 13, 1974. That night she was carrying damning safety documents with her, which of course disappeared from the scene. The FBI and the Oklahoma Highway Patrol predictably said her death was an accident and closed the case.

It was only thanks to the National Organization for Women that the Kerr-McGee Corporation was brought to trial. Facing imminent defeat in 1986, Kerr-McGee finally settled with Silkwood's family for $1.38 million.

In doing so, the corporation was really only settling for its responsibility in Silkwood's contamination at the plant. Its role in her contamination at home, and her murderers' identities, have never been uncovered.

Meryl Streep starred in the movie.

The Kyshtym Disaster, December 1957

The Kyshtym disaster represents one of the largest cover-ups the world has ever seen and easily wins the dubious honor of being the penultimate pre-Glasnost Soviet-style nuclear conspiracy.

The exact nature of this gigantic nuclear failure remains unknown, which is staggering in itself, because whatever it was contaminated 625 square miles of Soviet land, including 14 lakes, the industrial town of Kyshtym, and a full complement of 30 smaller communities with populations less than 2,000.

The Soviet government's solution to the problem was to censor all references to the surrounding events in any and all documents, including removing these towns' names from all maps made after 1958. The party line was that it never happened, and not only that, but there had never been anything there in the first place.

The conspiracy spread like a virus as the CIA found out about Kyshtym, and somehow the Atomic Energy Commission did too, but no information was ever allowed to be disclosed to the public.

Ralph Nader himself finally secured evidence via the Freedom of Information Act—the documentary proof that these tentacles of government knew about it all along and never mentioned it. Instead, they continued to sell the rest of the world on nuclear safety, and the safety of U.S. reactors in particular; yours for a low, low price.

Of course, there were many people in the area of Kyshtym at the time of the contamination, and their stories gradually filtered out to the international community through a channel of ex-Soviet scientists. Apparently, the residents in and around Kyshtym had all been shipped out at gunpoint and their houses burned down to prevent them from ever returning.

Several theories have been proposed by various researchers as to the nature of the Kyshtym disaster. All anyone knows for sure is that there was a plutonium production facility built around there in the 1940s, and over the years it suffered at least one, but possibly several, explosive nuclear and chemical accidents.

Some believe the major source of contamination was simply the massive pollution generated from the day-to-day operations of the facility, which destroyed the local river and generated radioactive acid rain over the entire area.

Now that the government responsible for the catastrophe is out of power, no one is accountable. All that's left is an old road sign along a dead stretch of road in the Ural Mountains: "Drivers are warned not to stop for the next 30 kilometers. Drive through at top speed with your windows up."

Recommended Reading: *Greenpeace Book of the Nuclear Age* by John May; *The Nuclear Barons* by Peter Pringle and James Spigelman

CHAPTER 28

Quantum Pollution

LONG ISLAND, NEW YORK, IS HOME TO ONE OF THE U.S. government's most advanced nuclear accelerators, the Relativistic Heavy Ion Collider (RHIC) built by Brookhaven National Laboratories. This device was the scene of a short-lived but extremely gonzo debate that began in July 1999. The topic? Whether the RHIC, the most powerful device of its kind in the world, could destroy the entire planet and leave behind no trace of it whatsoever.

The answer? Don't worry. They set up a committee of experts. And according to an RHIC scientist, the risk of total annihilation of the whole world is "astonishingly small."

Nuclear accelerators comprise some of the largest machines constructed by humankind. Often stretching for miles underground, their purpose is to accelerate subatomic particles around and around, smash them into each other, and record the grisly collisions on film to be scrupulously pored over, a veritable NASCAR racing of the physics world. The aim is to pry these particles open and scry their secret forces. What could go wrong?

In the case of the RHIC, some physicists felt that plenty could go wrong. The director of Brookhaven National Laboratories, John Marburger, felt these concerns were serious enough to address via a committee of physicists.

The committee's primary concern (the doomsday one) involved the possibility that the particles in this particular smashup derby could achieve a density high enough to produce a gravitational field strong enough to collapse into a mini black hole. This black hole would then fall to the center of the planet, eating matter the whole way down like Pac-man. Then, in a matter of something like five or six minutes, the entire world would be consumed.

The other concern addressed by the committee, while also assigned an extremely low probability, was apparently thought to be somewhat more likely than the outright doomsday scenario. In this scenario, the RHIC accidentally creates a new type of particle called a "strangelet," which chain-reacts uncontrollably and converts anything it touches into more strangelet matter. Sounds just as fatal as the doomsday one, doesn't it?

Our nagging doubt is that the odds may very well have been underestimated by the committee of physicists, all of whom were probably drooling over the chance to use the RHIC themselves, bright shiny toy that it is. More and more accelerators of greater and greater power are bound to be constructed in the 21st century—what new particles will be born in these machines, and what fundamental forces will be stretched ever thinner? Can "mishaps" with this technology be accurately foreseen? Could a black hole in New York or a "strangelet event" be contained, or cleaned up?

We do not wish to play Chicken Little here, but could we be seeing the emerging edge of an entirely new breed of environmental disaster, a kind of disaster that makes Chernobyl look like a bouquet of flowers?

Recommended Reading: "A Little Big Bang," *Scientific American*, March 1999; BNL press statement, July 1999.

CHAPTER 29

It's the Food Supply, Stupid

ATTENTION GENETIC ENGINEERS: YOU ARE ENDANGERING the food supply. You are committing the same errors in your thinking as those guys who brought rabbits to Australia, the guys who thought, "What could go wrong?"

Plenty—duh! Pollen drifts and contaminates non-engineered plants as they exchange genes. All the studies have shown contamination to be easier than first suspected. This "genetic pollution" is a spill that cannot be cleaned up. How far does pollen drift? The last time they checked, they found that it drifts way, way farther then expected. There is no containment. Hello. This is a serious problem. Genetically modified (GM) genes from drifting pollen get mixed up with non-GM genes and make GM plants, which spread more GM pollen much easier than expected. (This follows the same model as the Yucca Mountain All-Stars who loudly assured the world that Yucca Mountain was a solid piece of rock that makes a great nuclear container; water could only seep through the mountain in eons. But oops, turns out water pours through this leaky barrel of a mountain in only 40 years. Another example of science hijacked by government or industry or both.)

You don't have to be a genius to see that we need more than a precautionary principle. Ban GM crops before a genetic Exxon Valdez or Chernobyl disaster.

We are in complete philosophical agreement with those who criminally destroy GM crops under the cover of night (note to the Attorney General: we are not eco-terrorists and please consider that we have here stopped short of advocating any such criminal action).

The paper "Transgenic Contamination of Certified Seed Stocks" by Joe Cummins, professor emeritus of genetics, University of Western Ontario, states:

The extensive contamination of certified canola seed with transgenes…is staggering. The Canadian canola crop extends over some 5 million hectares, of which roughly 60% are planted with transgenic varieties. The extensive cross contamination by transgenic varieties could have been foreseen and predicted at the time field trials of transgenic crops were carried out. By now, it seems unlikely that transgene-free canola can be produced in western Canada. It is disturbing that [the Canadian Food Inspection Agency (CFIA), a sub department of Agriculture Canada] appears to be totally unconcerned over the extensive contamination, which is evidence of gross negligence in oversight on its part.

Professor Cummins is a member of Independent Science Panel. With dozens of members and representing seven countries, they are a cross-disciplined group with experts from disciplines such as agroecology, biomathematics, chemical medicine, microbial ecology, molecular genetics and virology, and many others. In their "final report" (posted at www.i-sis.org), they provide a summation of their reasons for supporting a GM crop ban, which we provide here in digest form.

1. GM crops failed to deliver promised benefits such as increase in yields or reduction in pesticide use.
2. GM crops pose escalating problems on farms, such as the super-weed problem.
3. Extensive transgenic contamination is unavoidable and is happening right now.
4. GM crops are not proven safe, and in fact, raise serious safety concerns.
5. Dangerous gene products are incorporated into food crops such as genes and fragments of genes that are known to suppress the immune system.

6. "Terminator" crops spread male-plant sterility via pollen.

7. Genetic engineering greatly enhances gene transfer and recombination, the main route by which super-viruses and bacteria are created. Newer techniques such as DNA shuffling allow the creation of millions of never-before-seen viruses in minutes. In fact, disease-causing viruses and bacteria are the predominant materials and tools of genetic engineering—just like in the making of bio-weapons. (Will we be able to tell if a genetic pollution event is terrorism or just commercial biotechnology?)

8. Bacteria in your gut may take up GM DNA from the GM food you consume. This could transfer traits like antibiotic resistance to pathogenic bacteria, making for difficult-to-treat infections. There is also the danger a cancer-triggering effect of having GM DNA jump into our genomes. This could even come from eating non-GM animals that have consumed GM food, since GM DNA can survive digestion.

9. Some GM constructs could be especially unstable and prone to gene transfer and recombination, with the attendant hazards of gene mutation, cancer, re-activation of dormant viruses, and generation of new viruses.

10. There has been a history of misrepresentation about GM and suppression of scientific evidence. Key experiments haven't been performed, or were performed badly and then misrepresented. Many experiments were not followed up, especially if a possible adverse health effect was indicated.

Get online and find out who your representatives are, and then write or call them and say, "I want you to ban genetically modified crops, and I vote. It's the food supply, stupid."

Recommended Reading: "Transgenic Contamination of Certified Seed Stocks" by Joe Cummins at http://www.i-sis.org.uk/TCCST.php

CHAPTER 30

Cold Fusion: Fire Your Physicist and Hire a Chemist

COLD FUSION IS ROOTED IN THE OTHERWISE RESPECTABLE science of electrochemistry, the study of electrically conductive solutions. This has been a legitimate field of study for chemists for more than 100 years. In electrochemistry, various metals (called electrodes) are combined with electrically conductive solutions (called electrolytes). When electricity is passed through these configurations (or "electrolytic cells"), the effects can be quite interesting and useful. For instance, this is where we get household batteries.

Electrochemistry is sometimes whimsically referred to as "two-dimensional chemistry," because the action in an electrolytic cell takes place purely on the surface of the electrified metal as it sits in the electrolyte solution. The surface of the metal is where the electricity (in the form of electrons) emerges into the electrolyte solution, and thus the surface of the metal is where the chemical reactions occur.

The practice of electrochemistry can be enormously complicated by very small changes to the surface of the electrode. Since the critical reactions are dependent on an essentially two-dimensional surface, any change to that surface can completely screw up the hoped-for reactions. A layer of impurity only one atom thick can drive an electrolytic reaction in the completely wrong direction. Electrochemistry is a very difficult, precise, and demanding branch of science.

But some very interesting effects can be achieved by the skilled practitioners of this discipline. For starters, some metals can be made to absorb enormous amounts of the various forms (or isotopes) of hydrogen. Put an electrified metal rod of, say, palladium, into an electrically conductive mixture of water and lithium. The hydrogen atoms and the oxygen atoms of the water will separate from each other, and, as the oxygen bubbles to the surface of the solution, the hydrogen will dissolve into the molecular lattice of the palladium. This effect has been known since 1870 and can be very useful—for instance, in separating hydrogen from other gasses.

The cold fusion part begins in the mid-1980s, when two of the world's top electrochemists, Martin Fleischmann and Stanley Pons, decided to see how far they could push this effect. They were intrigued by scattered hints in the electrochemical literature that something gonzo might happen if they loaded up a palladium lattice with isotopes of hydrogen and then kept loading. There had been similar attempts to do this in the 1920s that had never been followed up. It seemed as though electrically driving hydrogen into a palladium lattice could cause the hydrogen atoms to lose their electrons to the electric current in the metal. This would leave just the bare hydrogen nuclei hanging out in there, racking up tight like billiard balls. It really was not that crazy for these scientists to ask: How close can we get these hydrogen nuclei together? If we push them for months, can we...fuse them?

They gave it a shot. They assembled a roomful of electrolytic cells and kept them bubbling away for months. The whole time they meticulously measured every bit of heat produced by the cells, and when they were done they crunched the numbers. To their delight, they found that a little excess heat had been produced. This was a potentially marvelous discovery as it promised a clean new source of energy.

The rest is history. Rushed into a press conference by their university, the nuclear physicist community came down on them like a ton of bricks. Amazingly, the excess heat data were totally ignored and overshadowed by the physicists' demands for proof of nuclear by-products. Fleischmann and Pons, it turns out, had flubbed their nuclear by-products measurements, and were summarily condemned as crazy hucksters. Meanwhile, the excess heat measurements survived intact, and have been reproduced in labs across the world for more than 10 years, along with the much-vaunted nuclear by-products.

Cold fusion is a science vs. science story, pitting the nuclear physicists against the electrochemists. For whereas the excess heat of the cold-fusion effect can be pursued in a chemistry lab with ordinary equipment, the nuclear physicists have been trying for years to generate heat in multibillion-dollar "hot-fusion" reactors, and they still can't get it to work (fusion being different from fission, which works fine—if your definition of "fine" includes nuclear waste and chronic safety problems).

So the nuclear physicists are a little edgy that someone might muscle in on their territory from an entirely different discipline. They stamped out the cold-fusion claims before any of their hot-fusion funding got diverted. Spoiled on Manhattan Project-style excesses and cushy government contracts, the nuclear physicists huffed and puffed and blew the chemists' house down. Or did they? Without funding or recognition, it normally takes a fledgling science about 30 years to "make it." Watch this space.

CHAPTER 31

Interview with Cold Fusion Researcher Edmund Storms

COLD FUSION RESEARCH HAS QUIETLY PICKED UP STEAM across the globe, and many researchers claim to have duplicated Pons and Fleischmann's results in painstaking experiments. Edmund Storms is one member of this international community of privately funded researchers who studies cold fusion. We caught up with him in his home in Santa Fe. (Thanks to Dave "The Invisible Man" for helping to conduct this interview.)

GONZO SCIENCE: What is your background?

STORMS: I graduated with a Ph.D. from Washington University in St. Louis. There I studied radiochemistry, which is the use of radioactive elements to study the chemical environment. Ironically, I started in the very same subject that I'm in right now, although this application of radiochemistry is opposite to the conventional approach, that is the chemical environment is used to study nuclear processes. I took a job at the Los Alamos National Laboratory and worked there for my entire career—34 years— which included a lot of work on nuclear programs of various kinds.

I was working there when Pons and Fleischmann made their announcement. This announcement started one of the most exciting times at Los Alamos—for everybody, not just for me. Everybody there was enthralled with the idea that a nuclear reaction could be made to occur so simply.

GONZO SCIENCE: Can you explain Pons and Fleischmann's announcement at the University of Utah?

STORMS: They claimed to produce nuclear energy from a fusion reaction—that is, two deuterons fusing to make helium and various other products.

They proposed to make this process occur in an electrolytic cell, which is a device having the size of a small jar with two electrodes in it. Current is caused to pass between these two electrodes and through a solution containing heavy water (D_2O). The heavy water decomposes, causing deuterium to dissolve in the palladium electrode. Once enough deuterium has entered the palladium, the individual nuclei come together and fuse to make energy. By conventional thinking, this process is absolutely impossible under these conditions.

GONZO SCIENCE: Why do you think Pons and Fleischmann's early nuclear measurements were so full of errors, and do you think there's a case for fraud?

STORMS: They expected the reaction to produce neutrons, because they thought a typical hot fusion reaction would occur within their cell. They didn't have a method for detecting neutrons, so they looked for gamma rays that result from neutrons interacting with water surrounding their cell.

Not being nuclear physicists, they did not know how the gamma ray energy spectrum should look. So they accepted the energy spectrum as being close enough to expectations and they published the result. Someone who was knowledgeable about gamma ray spectra said, "Hey, wait a minute, this is wrong." So, they shifted their spectrum to correspond to how it should look. In other words, they corrected a calibration error in the measured energy, which is not an uncommon procedure. However, this is usually done before the results are published. As a result, they appeared to be lying about neutrons being present.

I think it was an honest mistake.

Part of the problem resulted because they were under extreme pressure from skeptics and crazy people. At that time, Pons and Fleischmann were absolutely inundated by the attention. They were getting hate mail and death threats. They were getting calls from people who wanted to know how to develop the method so that they could make millions, and they were being called by physicists who said they didn't know what they were talking about. They got calls in the middle of the night until they had to have their phone numbers changed, which resulted in peace for a few days. The press would not leave them alone. As a result, they were under intense pressure. In fact, their experience is a lesson to anyone who plans to announce an important discovery.

I think being totally open under such conditions would be very difficult. I don't think they used fraud. It was just sloppiness on their part and probably lack of sleep. Later, detailed examination of their work shows their claims for anomalous energy to be justified. Since then, hundreds of successful replications show that the claimed effect is real. However, skeptics did not wait for this evidence before rejecting the idea.

GONZO SCIENCE: Initially, what made you think that cold fusion was worth pursuing?

STORMS: My background involves materials science. Anybody who understands materials knows just how ignorant science really is. In addition, professors Pons and Fleischmann are highly respected and very knowledgeable scientists. These two facts caused me to give them the benefit of the doubt. A lot of people who are now skeptics were willing to give them the benefit of the doubt then as well. People did not say, "Oh, you're nuts and this is stupid." Instead, they said, "Hey, maybe you've got something worth exploring." The problem came when most people failed to reproduce the claimed results.

The few of us who were lucky to reproduce the results stayed in the field. The people who couldn't reproduce the claims decided that the people who were getting positive results must be deceived. Whenever they are asked about whether or not [cold fusion is] real, they say, "Of course not, it can't be because my friends and I can't make it happen."

GONZO SCIENCE: Did the Pons and Fleischmann announcement take your career in a different direction?

STORMS: Not immediately. I was part of a program to put a nuclear reactor in space. When Pons and Fleischmann made their announcement, people at Los Alamos dropped everything they were doing. Huge meetings were held, faxes went back and forth, and everybody was really excited. People tried all kinds of different experiments. I proposed an experiment that was thought reasonable, so the DOE gave us a quarter of a million dollars. We dropped what we were doing and started to work on this project.

The director of the lab saw me on the street and said, "This is the most exciting time since the war! Physicists and chemists are actually talking to each other." Physicists didn't know anything about palladium, so they had to ask a chemist. Of course, chemists didn't know anything about the nuclear processes, so they had to ask a physicist. Everyone was talking, which made the work very exciting.

Of the many attempts, only three succeeded. I was lucky to be part of one successful effort. Naturally, seeing is believing. Unfortunately, because most of the efforts failed, few believed the positive results.

It's strange to hear skeptics say that they require claims to be published in peer-reviewed journals for them to believe the results. All work that came out of Los Alamos was peer-reviewed by many people. I have never had a paper so thoroughly reviewed as the ones I published when I was working there. The work was reviewed far more thoroughly than [what] is done by a journal. Yet most scientists refused to believe the results. Instead they would believe negative results that were later shown to be poorly done and filled with errors.

GONZO SCIENCE: What was the most exciting moment in your lab?

STORMS: My most exciting moment was when a cell actually started making energy. I had constructed a calorimeter [a device used to measure heating power] and was sent a piece of palladium from Japan, where a similar piece was found to have made extra energy. The cell ran for about a week, doing absolutely nothing. I kept thinking, "Who needs this? This is just obviously nuts. Who could believe this could be real?"

One day I discovered that the line on the chart was starting to move away from zero. In other words, some extra energy was being made. I immediately thought something had gone wrong. So I checked everything, but nothing was wrong. The calorimeter was working perfectly. Meanwhile, the extra energy just kept increasing. Over the last 14 years, I have seen this effect happen dozens of times, but with enough failures to know that the effect is not caused by error.

GONZO SCIENCE: Have you had any bitter disappointments or major setbacks?

STORMS: I have been disappointed in the behavior of my fellow scientists, I have been disappointed in how the government treated this discovery, and I have been disappointed in the Patent Office because of their failure to examine the evidence objectively.

I have been impressed by a few individuals who have the money, the social conscience, and the intelligence to provide money to people in the field. Most work done in the U.S. is supported by a few individuals of extraordinary insight who see this phenomenon as a possible solution to the world's very real energy problems. A few of these people are in government and work at major laboratories. However, many more people have prevented work in the field and publication of what is known.

The field at the present time has some advantages to people like myself because significant progress can be made using very little money. If the government were supporting the field with multimillions of dollars, this wouldn't be true. I suppose I should thank the skeptics for keeping big money out of the field so that I can continue my work without too much competition.

My disappointments are more of a general nature, rather

than being personal. I haven't been treated as badly as some people. A few people in the field have paid a very high personal price—for example, John Bockris at Texas A&M. His life at the university was made very unpleasant by his fellow professors, even though the university is supposed to promote academic freedom. In addition, Pons and Fleischmann were treated very badly by their fellow scientists, causing Pons to move his family to France and to renounce his U.S. citizenship. Other people in the field have been trying to do their work with very little support and with great difficulty in getting their results published. Most people in the field have good reason to be bitter. In general, rejection has moved from the normal process expected in science to a personal, extreme level more typical of conflicts involving religion.

GONZO SCIENCE: As we understand it, MIT came out with a pronouncement to the official media, and that pretty much just stopped cold fusion research.

STORMS: Three laboratories helped stop cold fusion research. MIT, Caltech, and Harwell in England all helped inhibit research by publishing negative results that were later shown to be based on incorrect analyses. Those three laboratories tried to duplicate the effect early in the field's history without a full understanding of what they were doing. The failed attempts set the tone for later rejection.

GONZO SCIENCE: Was that intentional or unintentional?

STORMS: I think the negative results were unintentional in the sense that whenever a very small signal is seen, the natural reaction is to suspect an error and dismiss the result. And that's perfectly normal, reasonable, and rational. In addition, several of these early measurements made errors based on ignorance of the methods they were using. Instead of addressing the claims for the small signals being real, these people rejected the results as being error. Consequently, the claims were never properly examined or replicated.

Because of very strong pressure within the physics community, especially from the theoreticians, people were encouraged to dismiss the small effects. The claims flew in the face of what everybody knew was possible. They could not understand why or how the claimed fusion reaction could possibly work.

GONZO SCIENCE: So how can there be a consensus reality, or a consensus on anything?

STORMS: Theoreticians fight among themselves until one theory is more believed than another. If a claim is totally outside of any accepted theory, no respected theoretician is available to fight the theory battle for you. As a result, all theoreticians gang up against the idea. Fortunately that problem is slowly being solved for cold fusion. A number of theoreticians have taken an interest in cold fusion and have started to propose explanations. They have started to fight with conventional theoreticians by saying, "Now wait a minute, you have overlooked this and you forgot about that." The war of theory has now begun.

GONZO SCIENCE: The mechanism for the production of excess heat in cold fusion is as yet unknown. If you had to choose one theory from the bunch, which one would it be and why?

STORMS: Let me define that question a little more clearly. We know that energy is coming from a nuclear reaction. We know that one major nuclear reaction producing most of the energy is fusion, which produces helium. As a result, a mechanism for one of the many observed nuclear reactions can be proposed.

We don't know why or how that mechanism occurs in the environment where it is found to operate. These nuclear reactions don't occur in every environment. The environment in which they do occur is obviously very rare. The question is, what makes this environment so unique? We don't yet know the answer to this question.

Now, when you ask about which theory or model or mechanism I most believe, it has to be one that focuses on the unique environment. Two mechanisms appeal to me at this point.

Under certain conditions, the deuterons are proposed to form waves, which interact without a Coulomb barrier. The resulting energy of interaction is dissipated as small packets into the surrounding lattice. As a result, the combined waves slowly convert to helium. That theory solves a number of problems. Of course, a person has to believe that particle-wave conversion is possible in such an environment and at room temperature, which is not an easy thing to believe. Once this process is accepted, the theory becomes very attractive.

The other theory involves the surrounding electrons, which have the ability to coalesce and form a coherent structure. Electrons do this in superconductors. There they form a structure consisting of electron pairs, called Cooper pairs. It's conceivable that under other conditions they might form Cooper six-packs or even more complex structures. Such a coherent structure might be capable of intervening and neutralizing the Coulomb barrier, thus allowing the various nuclei to get close enough to interact.

The literature contains more than 500 papers describing themes on a variation and variations on a theme of many mechanisms. Lots of possibilities have been exported.

GONZO SCIENCE: Back to your career. How long have you worked on cold fusion?

STORMS: I worked on cold fusion at Los Alamos from 1989 until I retired in 1991. After I left Los Alamos, my wife and I constructed a home in Santa Fe, next to which we built a laboratory where I'm currently doing research.

GONZO SCIENCE: So 14 years later you're still working on cold fusion and on developing devices?

STORMS: Yes. I have focused mainly on the Pons-Fleischmann method, that is, the electrolytic approach. Since Pons and Fleischmann's announcement, 11 other techniques have been discovered to produce the same result, both nuclear products and/or excess energy. It turns out that the Pons-Fleischmann technique is one of the more difficult ones to replicate.

The Pons and Fleischmann method is difficult because it relies on the use of palladium. Palladium having the necessary ability to support a high concentration of deuterium is very rare. Once this problem was realized, I started making my own electrodes out of finely divided palladium. This approach has improved reproducibility. Consequently, the effect is easy to reproduce once the right material is used. People are starting to explore other materials and having success.

GONZO SCIENCE: Have you built any devices?

STORMS: I've built many devices in order to better measure heat production. However, we're not at a point where commercial devices can be constructed because the amount of energy being generated is still rather small. However, it's very clear that only a small amount of material is active, which results in a very high power density. In other words, only a little bit of material is producing all of the energy. Unfortunately, not much active material is normally present. The challenge is to find ways to make more of this active material so that the resulting power is increased to commercial levels.

GONZO SCIENCE: Would you venture a prediction about how long it will take to bring cold fusion devices to market? Are we going to have to wait a whole generation?

STORMS: That very much depends on two things. It depends upon how lucky somebody might be in finding ways to make the active material. It also depends upon the attitude of the establishment and whether or not they will apply significant money to the studies. For example, if the DOE or the government or somebody starts funding it at the multi-million dollar level, we could have a practical device in a few years. If the government behaves as it has in the past, it might be 20 or 30 years before anything useful is seen, unless it is developed in Japan. Somebody in another country might hit upon the secret ingredient. Should this happen, that country will control the phenomenon. They will develop it and they will market the products to the rest of the world. Should that happen, the United States would be in serious trouble. Besides OPEC controlling our oil, we would see Japan or another country controlling our cold fusion energy.

GONZO SCIENCE: It's no surprise that the oil industry, and the nuclear industry in its current form, are perhaps trying to stifle cold fusion research, because it would cut into their bread and butter practically overnight.

STORMS: I don't think they're working very hard to stop cold fusion. If I were employed by the oil industry, I might conclude that maybe someday this energy source might be important, but oil is important right now. Besides, most of the physicists in the world are fighting my battle for me. And I don't have to pay them, I don't have to convince them, I don't have to do anything! I can sit back and let them reject the idea for me. They will effectively protect my business. If somebody should make the process work at high levels and start manufacturing devices, then I will get worried. But at that time, I can buy into the field and continue to make money selling cold fusion devices.

GONZO SCIENCE: Your results indicate that cold fusion is real, and yet many people consider cold fusion to be dead. What happened?

STORMS: The first problem was to reproduce the claims. Science absolutely requires, and reasonably so, that other people be able to duplicate what is claimed. Initially that was very difficult to do because Pons and Fleischmann really didn't tell people exactly what they had done. To a large extent, they didn't even know the important details. They happened to acquire some good palladium, but they didn't know why it was good. Gradually, people understood the necessary properties of palladium that made it active. Such material is still difficult to find.

I can give you a little anecdote of my own. The Japanese were involved in cold fusion from the beginning. An investigator in Japan by the name of Professor A. Takahashi ordered palladium from Tanaka Metals, a metal company in Japan. They supplied a plate of palladium that he cut it into suitable pieces. These were sent to various people, including myself, and he studied a piece. I found the piece to make extra energy. A scientist in Italy as well as Takahashi himself in Japan also measured extra energy from their pieces. As a result, three independent duplications of the Pons and Fleischmann claim were reported, yet ignored.

Eventually, Professor Takahashi ran out of the good material. So he asked Tanaka Metals to make another batch, which they did. He sent a piece of this new metal to me. Unlike the first material, this palladium was dead. Fortunately, I had made measurements that showed why this difference occurred. The first piece of palladium did not crack. As a result, it was able to retain the high deuterium concentration that is required. The second piece cracked like crazy, so the deuterium leaked out. It was like trying to fill a leaky bucket.

We told Professor Takahashi about our experience and Tanaka Metals was contacted once again. Apparently, they claimed to have made the second batch exactly like the first, but they hadn't done it quite exactly. They had made a few little changes. So they made a third batch exactly like the first, without any changes. This material was found to make energy. It didn't make as much as the first batch and it cracked a little more, but it still made energy. So here is an example of how sensitive the effect is to the nature of the palladium being used. Tanaka Metals didn't know why one batch worked and the other batch didn't. They just knew they had made a few little changes that they thought were insignificant. It turns out these small changes made all the difference in the world.

Unfortunately, palladium is not made using the same process as before. Unfortunately, material is not being made for our use. Instead, it is made for the jewelry and catalytic industries. The refiners aren't interested in what we do with the metal. As a result, it's very difficult now to find palladium that has the necessary characteristics. That's why replication has been so difficult. Therefore, a person needs to make [his or her] own palladium, so to speak, by depositing it on other materials using electroplating or by vapor deposition, which I now do.

The Russians, French, and Japanese are exploring various

techniques. A big program is ongoing in China. The Italians have a large program. The Japanese have started the first official cold fusion society. All over the world, people are seeing these phenomena. A problem remains to explain the process. How can the nuclear reactions be made to happen at a sufficient rate so that useful energy results? That's the challenge.

GONZO SCIENCE: The U.S. Patent Office will not issue patents for cold fusion technology. Isn't that unscientific?

STORMS: It's totally unscientific. It's arbitrary, it's capricious, and it's completely wrong. There's no excuse at this point for that kind of attitude. Such caution might be excused right after the announcement, but now a large amount of supporting evidence has been reported from laboratories all over the world. Nearly one hundred replications have been reported, well over a hundred peer-reviewed papers on the subject have been published, and a thousand references describing the effect are available for study. Ten international cold fusion conferences have been held, along with papers presented at APS, ACS, and ANS meetings. All of this information is available to the patent examiners. They have chosen to ignore the evidence in violation of their duty.

GONZO SCIENCE: Who is creating the agenda that the Patent Office is following?

STORMS: This rejection results because many scientists sincerely think the observations are impossible. When somebody from the Patent Office asks an expert, the impossibility of the effect is described and the embarrassment of supporting such work with a patent is emphasized. Unfortunately, these "experts" are totally ignorant about what has been discovered because they have not taken the time to read published results.

GONZO SCIENCE: What is your perspective on the idea that the nuclear physicist community is spoiled on Manhattan Project-style government largesse, and that cold fusion was quashed because it was perceived as a threat to the lavish funding for hot fusion research?

STORMS: There's no doubt in my mind that control of science has shifted over the years and is now largely in the hands of the physics community. That shift occurred largely because of the successful atomic bomb program. Also, theory of nuclear interaction at high energy is highly developed and is considered to be the only understanding that can apply to nuclear interaction. That kind of success gives people power to influence the direction of research and confidence that they know all the answers. In earlier times, power was in the hands of chemists because of their success in creating all kinds of chemical products—plastics, dyes, pharmaceuticals, and so forth.

I don't think this shift in influence is all bad. Only when individuals use this power without the humility science demands does a problem arise. Scientists must be and should be humble in the face of all that we do not know about nature. We know from history that mankind's understanding changes dramatically. What is thought to be advanced today will seem primitive in a hundred years. Today, we know practically nothing. We have only the slightest understanding of what nature is really like. Unfortunately, some physicists think they know everything. They think they have nature figured out. That kind of arrogance stands in the way of progress.

GONZO SCIENCE: We understand that demonstrations have been done for the DOE with strontium-90, whereby radioactivity has been remediated. Is that correct?

STORMS: I know of some people who are working on remediation of radioactivity, which describes the ability to reduce the radioactivity of substances. Strontium-90 would be an example of an element to which the process could be applied.

This process tries to solve a problem we have created using ordinary fission reactors to make energy. Such reactors make large amounts of radioactivity that continues to accumulate. This dangerous waste could be eliminated if the unique environment used for cold fusion could be used to accelerate radioactive decay or transform these radioactive isotopes into stable isotopes. Such a process would eliminate a problem having considerable importance.

GONZO SCIENCE: Transmutation occurs, right? There are different elements that fall out?

STORMS: That's right. Once the unique environment is created, fusion, transmutation, and fission reactions are found to occur. The potential to change the nature of radioactive materials exists.

So far, people have been changing ordinary, stable materials into other stable materials. Radioactive materials can be changed as well. The potential exists to change any element into any other element. We're talking about true alchemy. Of course, that word sets physicists' teeth on edge.

GONZO SCIENCE: It's like we're talking about witchcraft. It's supposed to be forbidden.

STORMS: Witchcraft, perpetual motion, and cold fusion—they all seem to be in the same category in many people's minds. However, each provides a window into aspects of nature that are yet unknown. The challenge of science should be to reject myths and accept what is real, without rejecting the whole collection of information. After all, most of what we take for granted today would seem like magic to a person one hundred years ago.

GONZO SCIENCE: Is there a "cold fusion underground" at Los Alamos?

STORMS: Some people at LANL [Los Alamos National Laboratory] are sympathetic to the reality of the phenomenon. Unfortunately, the DOE has officially concluded that cold fusion is not real. Because the DOE funds the laboratory, the DOE forbids spending any money to study the subject. A few private laboratories, like my own, are working on the problem and several universities, for example the University of Illinois and the University of Portland, have programs. Much more work is being done in other countries, some supported by governments, universities, and/or corporations.

GONZO SCIENCE: So where are you right now in your research?

STORMS: I have gone from disillusionment with the Pons-Fleischmann effect to a study of the Case effect, i.e. gas loading of finely divided palladium on carbon. I had great hope for this study, but alas it proved to be too difficult to reproduce. As a result, I returned to the Pons-Fleischmann effect, but with some new ideas of how to proceed. I feel confident now that this method can be made 100 percent reproducible.

GONZO SCIENCE: So you're still networked into the cold fusion community of the world? And you're contributing, you're exchanging information?

STORMS: Yes, I'm still deeply involved in the subject. I have written a series of review papers, continue to publish my experimental results where I can, and have helped create the www.LENR-CANR.org website. All of my publications are available there as pdf text files.

GONZO SCIENCE: What do you foresee as the future of cold fusion?

STORMS: Sooner or later somebody will figure out how to make this work in a very efficient way. We will then have in the corner of every house something the size of a small refrigerator. This will supply all of the heat and electricity for the house. Being hooked to the electrical grid or to the gas line will no longer be necessary. The transition will start slowly, probably mainly in the Third World. As yet, we don't know how to make the nuclear reactions occur at very high temperatures, so it's hard to turn the heat into electricity. However, space heating is possible and may be the initial application.

    The nuclear generator will give heat whenever it is needed. After all the fuel is used up—maybe after a year or two—the active part will be easily replaced.

    I have no doubt that sooner or later cold fusion will supply energy very cheaply, without any pollution whatsoever, and in a sustained way. One part in 6,000 is deuterium in all water so a huge amount of deuterium is available. Fusion using the cold fusion process, in contrast to the hot fusion process, produces no bad products, no CO_2, and no radioactivity. The method is safe and I believe it will eventually be very convenient and cheap to use.

GONZO SCIENCE: The debate over cold fusion strikes us as a science vs. science story. In some respects it is the electrochemists vs. the nuclear physicists. What do the electrochemists know that the nuclear physicists do not?

STORMS: That's a very good question. Nuclear physicists are handicapped by being trained to understand nuclear interaction using high energy or neutrons. This training makes them reluctant to consider any other method to overcome the Coulomb barrier. The cold fusion process involves an entirely different mechanism, which is very different from conventional experience. So, scientists are hard pressed to judge the reality of the proposed mechanism. Scientists normally judge the reality of claims based on their ability to reproduce the effects and on their ability to explain the effects. So far, the effects have been difficult, but not impossible to replicate and the explanations have not been very satisfactory. These limitations have slowed acceptance.

In addition to these rational reasons, most scientists are so sure the claims are nonsense that they don't bother to read about the new results. As a result, the new, impressive information is ignored so that their attitude remains unchanged. The absence of the new information in respected scientific journals and in the press does not help change this attitude

GONZO SCIENCE: Have you ever faced off with critics of cold fusion, and what are such exchanges like?

STORMS: Yes I have, and such interactions are very frustrating. It's similar to a discussion between a Muslim and a Christian about which view of Allah or God is correct. You get absolutely nowhere. If they act like gentlemen and they know you, they will hear you out. On the other hand, you might not get a word in edgewise. It's very frustrating. The discussion usually results in no solution whatsoever. In contrast, many people in the general public remember the Pons-Fleischmann announcement and their claims. These people are amazed that the effect has been so widely demonstrated, in contrast to what they have been told by the press. They generally come away from such a discussion with considerable anger about being so deceived.

Recommended Reading: *The Greenpeace Book of the Nuclear Age* by John May; *Excess Heat: Why Cold Fusion Research Prevailed* by Charles Beaudette

PART 8:
Altered States of Consciousness

CHAPTER 32

The International Conference on Altered States of Consciousness

THE 2001 INTERNATIONAL CONFERENCE ON ALTERED States of Consciousness grappled with the question of the scientific importance of altered states, and it did so in the face of almost certain derision by scientific skeptics.

Among the more than 40 presenters who converged in Albuquerque during November 2001 were top researchers and practitioners from all over the world. Gathered under the banner of "Enlightenment, Entheogens, Shamanism, and Peak Experiences" were people with backgrounds in the study of non-ordinary reality. They came from Brazil, Chile, England, France, Mexico, Nigeria, Slovenia, South Africa, and the United States, and many indigenous traditions were represented.

The mission statement of the conference was declared in its brochure:

> People are having experiences that are not explained by the old scientific paradigm, and we are called to reconsider our cultural view of reality. Each of us has experienced an altered state of consciousness at some time in our life. Sometimes it happens on its own, and comes to us as a surprise and sometimes we seek it out, through ingesting a substance or years of diligent discipline. It is at the root of every spiritual and religious tradition...

From shamanism to scientific investigation, the desire to understand altered states is universal.

We asked Barbara Gordon, one of the conference organizers, how she would respond to the charge that this was merely another "touchy-feely" New Age conference, not a scientifically grounded exploration. She responded: "We've actually tried to be really careful not to present it as a 'New Age' event. [New Age thought] has its place, but it's not where we're coming from. We started out doing the 'Science and Consciousness Conference,' and we really looked for people with academic credentials, but who were maybe doing things that were kind of pushing the envelope."

Indeed, the list of presenters at the conference was heavily weighted with a dozen Ph.D.s and at least one M.D. Various other world-renowned authors, thinkers, and agitators for consciousness exploration rounded out the roster.

One featured presenter was Skip Atwater, a ten-year veteran of the military's research using psychics for intelligence gathering. Atwater (author of *Captain of My Ship, Master of My Soul*) is now the research head at the Monroe Institute. His current gonzo work involves an audio technology called "Hemi-Sync," which facilitates training people to move through different altered states.

The conference also boasted an impressive list of sponsors which, while not directly from the comfortable middle of the scientific mainstream, nevertheless includes such respected names as the Institute of Noetic Sciences, the Botanical Preservation Corps, the Society for Scientific Exploration, the Association for Humanistic Psychology, and Temple University's Center for Frontier Sciences, among many others.

What is all the excitement about? What are some of the pressing scientific questions surrounding the issue of altered states of consciousness?

Science and religion eye each other warily as they circle around the topic of non-ordinary awareness. As well they should, for the perspectives of both science and religion seem to be uncomfortably, hopelessly intertwined when it comes to this complex of issues. And whereas some presenters at the conference attempted to tease these threads apart, some attempted to show how science and religion naturally interpenetrate within the altered states topic.

The umbrella term "altered states" encompasses such scientific hot potatoes as experiences where the dualistic categories of "subjective" and "objective" break down. The Eastern traditions bring "enlightenment" to the discussion, drawing upon a wealth of information about the navigation of inner space. Included too are those relationships to the natural order that can seemingly cause synchronicities.

Then there are the ritual/ceremonial techniques and technologies for accessing altered states which have been kept intact by indigenous cultures. "Shamanism," which received lots of scrutiny at the conference, refers to the techniques of ecstasy, dialogue with the spiritual world, and healing that are found nearly universally among indigenous people. "Entheogens," those plant and synthetic substances that by definition "bring forth god from within" (such as ayahuasca and DMT, both members of the tryptamine family) also drew much attention.

Also on the table for debate in this context are latent abilities, higher capacities for information usage and storage, unused capacities, and spontaneous healing.

And lastly, "peak experiences," which may be the least controversial and most widely recognized area of interest addressed at the conference. Sports are full of accounts of "being in the zone," that is, of experiencing a kind of spontaneous genius fueled by intuition and teamwork.

Some altered states reflect only a slight movement away from ordinary awareness, like simply quieting one's mind or hovering in that liminal state between wakefulness and sleep.

But then there are states so intense as to be overwhelming. In trance possession, for instance, the experiencer is said to become a temporary host to a descending spirit. And like the techniques of trance possession, the high produced by the tryptamine hallucinogen family reliably produces experiences of contact with other beings, those intelligences that the late ethnopharmacologist Terence McKenna termed "the Other."

Science has never encountered a more vexing subject than how to handle the accounts of those who enter altered states with plant sacraments and their derivatives. The scientifically relevant point about tryptamines is that, in terms of qualitative analysis from a community of psychedelics researchers, these substances reliably cause the experience of being projected into another space in which non-human entities are encountered. Both trance possession and the tryptamine hallucinogen high are linked together by their respective claims of contact with such discarnate entities.

Are the entities mere hallucinations? Some suggest that these entities represent autonomous parts of the psyche, loosed on the one hand by the drugs, and on the other hand by the rituals and ceremonies of trance possession. The most scientifically vexing possibility is that the experiences provoked by both the tryptamines and the techniques of trance possession are actually real.

Are the trance possession experiencers merely filling a cultural need? For some perspective on this question we turned to Nigerian master drummer Onye Onyemaechi. Onyemaechi is in a unique position to explain some of the cultural context surrounding possession as well as the techniques necessary to achieve this altered state. Onyemaechi explained that whereas a profound cultural shift has taken place across much of Africa over the last few generations, the phenomenon of trance possession remains.

For instance, as Onyemaechi describes it, his own grandparents never knew Christianity, and in their indigenous religion, "You have people dancing around the fire, or just dancing and drumming together, and chanting and singing…When that activity is heightened, then the deity of that particular village comes…into the body of the person that is possessed. And this person would then prophesize in these hallowed rites…and those around are also equally affected or 'trance induced' in the same manner; probably not as in-depth as the person that had been possessed, but in most cases, they are influenced by this experience taking place. So, those are the cultural aspects of it in that religious sense."

Onyemaechi goes on to describe how the phenomenon of trance possession stubbornly endures: "Now you have the Christians, [and] the religions that are now Christian; they no longer wanted to practice the traditional English Christian worship, like the Baptists and the Methodists and so forth and so on. They would like to create their own church called the African Church...so they would create their own church where they have the flexibility and the opportunity to worship god and praise god in ways that touch people and link them to divinity, direct...In these spiritual worship practices, the drums are used as well as the integration of the biblical text...So church begins. So through the time of worship: praises and songs, drumming and dancing. During that time, people are possessed. And they'll term it 'in the holy spirit.' You're not possessed by the deity of the village, because you're now in a whole different context...Usually when you're possessed you are seeing the future, you are telling about the future of what is going to happen for those members of the group at that particular time."

In these African-based religions, the descending spirit displaces the personality of the host to the point where they have no memory of the possession. Is it fair to ask if this is the same as the "lost time" of the UFO abductee? It makes sense to consider a connection here, as alleged "UFO entities" are also reported to deliver "prophecies" like the deities of the trance possession state.

Inasmuch as this connection is valid, and inasmuch as the possession experience and the tryptamine high lie along the same continuum, we might do well to ask: are trance possession, the tryptamine high, and other reports of entities describing the same "other" as viewed from different angles?

Why should there be congruence among such widely different types of experiences unless there is some underlying connection?

Does this connection merely reflect the topology of deep neurological structures, or something far more gonzo?

Recommended Reading: Bizspirit.com; anything by Terence McKenna, particularly *True Hallucinations*; *Food of the Gods*; or *The Invisible Landscape* (with Dennis McKenna). See also *Ecstasy: The MDMA Story* by Bruce Eisner.

CHAPTER 33

In Memoriam: McKenna, Sagan, Burroughs

In Memoriam: Terence McKenna

The Tim Leary figure of the 90s, Terence McKenna, died April 3, 2000. He was the world's premier ethnopharmacologist, which is to say he specialized in exotic drugs. The most interesting of these drugs—DMT, which is synthesized from various exotic plants—was McKenna's favorite. In our view, its chronic use is probably what killed him.

McKenna was a fearlessly creative thinker. His contributions to science include the theory that psychedelic mushrooms spurred humans to develop consciousness and language; he also formed the "Timewave" theory, described by chaos pioneer Ralph Abraham as "the first model for history that significantly transcends that of the ancients." The Timewave theory was developed in collaboration with his brother Dennis and some very weird experiences on dangerous drugs. The theory postulates that the hexagrams of the I-Ching are a kind of DNA of time, in which past and future exist simultaneously in a holographic fractal matrix.

Be that as it may, McKenna's *most* radical idea was that the ecstatic visions of the DMT psychedelic experience are, in fact, real. He seemed to be saying that the things you see on DMT are so exponentially weird that the best explanation for them is that they are really happening.

Usually smoked, the effect of DMT is immediate. Three hits push you down the rabbit hole into a comprehensive psychedelic morass. Then—and this is the widely reported detail of the DMT experience which consumed McKenna's genius—some of the nodal swirly bits of psychedelia break off and dance around you like munchkins trying to talk to you.

Aside from the drug's potency and rapid onset, this powerful impression of meeting otherworldly beings is the singular feature of the DMT experience. These powerful drug experiences caused McKenna to question the assumption that what one sees on DMT are merely chemically induced hallucinations. Toward the goal of gathering more data, McKenna took what he called "heroic doses" of DMT, often while peaking on several grams of mushrooms. These fact-finding tours of Wonderland continued throughout his adult life.

He began having seizures one day and was taken to the hospital, whereupon a diagnosis was made. McKenna had frontal-lobe brain cancer.

The frontal lobes are of course where all the action happens, and the precise area McKenna kept lit up like Las Vegas.

Was it worth it, Terence?

Of course, the answer is no. It was a tragedy. Quite literally, McKenna destroyed his mind with drugs.

He spent his final months undergoing a rash of incredibly expensive, invasive surgeries, gene therapy, all of it. McKenna surely didn't count on his adventures ending quite like this. Would he have done it the same way if he had known what it would so swiftly do to him?

Perhaps McKenna can be charitably viewed as a kind of self-sacrifice. There is, after all, a chance he was right about the slippery ontological nature of the DMT high. We should think of his death, in that case, not as a drug casualty, but more along the lines of an explorer or pioneer's death under extreme conditions.

An explorer opens up dangerous territories so no one else has to. If they make it back, we get to hear about it. Terence dazzled us with his traveler's tales of meeting the "self-transforming machine elves of hyperspace." These dispatches, and his intensely original cognitive style, are missed by his many followers, whose challenge now is to figure out what to do with it all.

Care must be taken, however, not to idolize McKenna. Questions linger about some moments in his career.

For instance, the UFO he reported seeing in the Amazon was almost certainly a hallucination or vision induced by drugs and fatigue. While he must have realized this, he chose to write about the experience as if it were an actual sighting of a UFO. One might argue that this can be naturally understood as part of McKenna's program to question the boundaries between internal and external phenomena. However, those waters are sufficiently muddy that this apparent instance of sleight-of-hand serves no useful purpose apart from self-mythologizing.

It is also reasonable to question the amazing coincidence of the "end date" of McKenna's Timewave theory falling on the same end-date as the Mayan calendar, of which he claimed to have no prior knowledge. Of course, the final test of the Timewave theory will have to wait until that very day in 2012.

The sad irony is that he very much wanted to live at least that long, but the selfsame drugs which helped spawn that desire, and the far-out Timewave theory itself, are what put him under the scalpel.

Ultimately, though, we should come away inspired by Terence; if for nothing else, let it be his untamed creativity, and his outstanding ability to view things in the freshest possible perspective.

He also fronted a rave band that wasn't too bad.

Carl Sagan, Buzzhound

Carl Sagan is America's own scientist-hero. His TV series *Cosmos* instilled in us an enduring love of astronomy and all science. During the cold war, his theory of nuclear winter arguably brought us all a step back from the brink of trading nuclear punches with the Russians. Carl Sagan personally showed how a man of science can change the world.

Carl Sagan, the people's scientist. Carl Sagan, popularizer of critical thinking and of a healthy skepticism. Carl Sagan, pothead.

After he died, Sagan biographer Keay Davidson outed him as a lifelong stoner. Carl, it seems, had been the long-time partner of one of the major players in the National Organization for the Reform of Marijuana Laws (NORML). Not only that, but in 1971 Sagan wrote an anonymous essay in a pro-pot anthology (*Reconsidering Marijuana*, edited by Lester Grinspoon) in which he credited the insidious weed with inspiring many of his scientific insights.

This revelation about Sagan's recreational drug use (we might say his inspirational drug use) has busted open a few stereotypes. Contrary to the pushers of zero tolerance, stoners can be smart; world-renowned scientists can get high; and cannabis won't wreck one's productivity. In fact, it may even increase one's productivity.

A lot of people are in a dither about this. Society is biased against any altering of one's neurochemistry whatsoever, like it soils you somehow—unless it's with caffeine, alcohol, or antidepressants, of course.

Interestingly, although a hero to the masses, Sagan was always a controversial figure among some of his peers. For instance, his raw scientific prowess was criticized by those who looked hard at his shoddy biology work and his half-assed mathematical abilities.

We might very well attribute a certain degree of Sagan's intellectual sloppiness to his predilection for smoking fat joints.

However, Sagan's real genius lies somewhere else. He was not a kind of human computer, like, say, Wolfgang Pauli is said to have been. But Sagan could rightly be said to be the godfather of exobiology, or the study of nonterrestrial life. He earned this impressive credit not by performing years of exacting calculations, but by fomenting an exchange of ideas between scientists in previously unrelated fields, such as biology and astronomy. Even as a student he was bringing together different disciplines.

Sagan's strongest, most productive methodological strategy was one of intellectual cross-pollination—thinking "outside the box" and encouraging others to do the same. This kind of "associative" thinking, as opposed to a linear by-the-numbers style, is precisely the kind of thinking that the cannabis high is known to encourage.

Sunday morning bong hits might very well have dulled Wolfgang Pauli's razor edge. But in Carl Sagan, we positively see how cannabis helped a thinker to break new ground.

If Carl Sagan could be said to have improved society—and not just society, but all of science—then the same could be said for cannabis. Irie, Carl!

The Science of William S. Burroughs

William S. Burroughs was one of the great writers of the twentieth century. He had an admirable ability to be entirely original in nearly every aspect of his life and writing. Ostensibly a novelist, Burroughs in fact wrote all kinds of prose, and his life's work includes much visual art as well.

An especially remarkable aspect of Burroughs' work is his take on scientific matters. He was well read in science and was fascinated with both Mayan and Egyptian hieroglyphics; he almost became an archaeologist. Mayan and Egyptian settings and characters were routinely featured in his books.

Burroughs generated a well-developed alternative scientific worldview that carried over into his fiction. As well as studying hieroglyphics, he became deeply involved in the orgone physics of Wilhelm Reich and, briefly, the sci-fi psychology of the Scientologists. Burroughs has been described as a "proto-ecologist" for writing prescient scenes of military-industrial ecocide way before it was cool. In addition, Burroughs was fascinated with virus theory and the evolution of language, and even postulated a link between the two.

One of Burroughs' overriding concerns was escape from linear time. He regarded the perception of linear time as a kind of conceptual prison, and he considered his writing to be a kind of nail file hidden in a cake.

Burroughs reasoned that temporal linearity was ultimately an illusion. Like many Eastern mystics and quantum mechanics, Burroughs thought that a "more real" world lay beyond the mechanistic model of linear time and its handmaid, the either/or duality. His life's work was an attempt to subvert the mechanistic, linear, dualistic model of the world, and to replace it with one that acknowledged a kind of holistic interconnectivity.

One of the tools Burroughs used to monkey wrench linearity was the so-called cut-up method. Applied to writing, the cut-up was essentially writing as montage—literally cutting up and rearranging blocks of text to derive new words, sentences, and associations. But in keeping with his unwillingness to distinguish between art and life, Burroughs took the cut-up a step further and employed it as a kind of magical weapon.

Since cutting up a two-dimensional page of words could sever and rearrange meanings and events, Burroughs thought it would be worth experimenting with the effect in real life by using tape recorders and photography.

Burroughs immediately realized that unusual juxtapositions of sound and image could function at least in part as a tool for psychological deconditioning, or deprogramming—a realization no doubt informed by his experience with the deconditioning methods of Scientology. He imagined that sound-image montages might be of use in brainwashing, for instance.

But consistent with his tendency to take ideas to their ultimate extent, Burroughs theorized that careful experiments might reveal an effect larger than a merely psychological one. Burroughs was essentially trying to rend space and time. The biography *William Burroughs: A Portrait* by Barry Miles contains anecdotes of Burroughs' audio-visual terrorism:

> [Burroughs said] "As soon as you start recording situations and playing them back on the street you are creating a new reality. When you play back a street recording, people think they're hearing real street sounds and they're not. You're tampering with their actual reality." [Burroughs] found that by making recordings in or near someone's premises, then playing them back and taking pictures, various sorts of trouble occurred. He immediately set out to exploit his discovery.
>
> "I have frequently observed that this simple operation—making recordings and taking pictures of some location you wish to discommode or destroy, then playing recordings back and taking more pictures—will result in

accidents, fires, removals, especially the last. The target moves." By 1972 [Burroughs] decided that his dissatisfaction with the Scientologists merited an attack on their headquarters. [Burroughs] carried out a tape and photo operation against the Scientology Center at 37 Fitzroy Street, in London, and sure enough, in a couple of months they moved to 68 Tottenham Court Road. The operation carried out there did not work and they still occupy the building.

The best example was an operation carried out against the Moka Bar at 29 Frith Street, London, W1, beginning on August 3, 1972. The reason for the operation was "outrageous and unprovoked discourtesy and poisonous cheesecake." [Burroughs] closed in on the Moka Bar, his tape recorder running, his camera snapping away. He stood around outside so the proprietor could see him. "They are seething in there. The horrible old proprietor, his frizzy-haired wife and slack-jawed son, the snarling counterman. I have them and they know it."

[Burroughs] played the tapes back a number of times outside the Moka Bar and took even more photographs. Their business fell off and they kept shorter and shorter hours. On October 30, 1972, the Moka Bar closed and the premises were taken over, appropriately, by the Queen's Snack Bar (p. 75).

Recommended Reading on McKenna: Anything by Terence McKenna, particularly *True Hallucinations*; *Food of the Gods*; or *The Invisible Landscape* (with Dennis McKenna)

Recommended Reading on Sagan: There's an assessment of Sagan's reputation among scientists in *The Nemesis Affair* by David Raup. Sagan's poor scientific prowess is detailed in *Carl Sagan and*

Immanuel Velikovsky by Charles Ginenthal. On the other hand, Sagan's books, and his positive influence on society, are everywhere.

Recommended Reading on Burroughs: All by William S. Burroughs: *Junky; Naked Lunch; Nova Express; The Wild Boys; The Job; Exterminator!; The Third Mind* (with Brion Gysin); the *Cities of the Red Night* Trilogy; *The Burroughs File.* See also *Literary Outlaw* by Ted Morgan and *The Portable Beat Reader* ed. by Ann Charters

CHAPTER 34

Interview with Remote Viewing Expert Paul Smith

REMOTE VIEWING REFERS TO THE MILITARY'S "PSYCHIC spying" program known as Project STAR GATE. It is supposed to be a kind of clairvoyance technique, and was employed by the government for more than two decades. It is still practiced and taught today by some of the graduates of the military program. Remote viewing is allegedly a skill that most people can learn that enables them to receive mental impressions of a distant scene, person, or event without involving their usual five senses. Targets are selected beforehand by higher-ups but are not known to the remote viewers themselves in order to guard against leading them into preconceived results. The idea was to psychically (as it were) access information from a distance. The ultimate efficacy of the remote viewing program has been a point of contention ever since.

Bouncing from practically every branch of the military and passing through academic labs like Stanford Research Institute (SRI), the remote viewing project left controversy in its wake.

We got in touch with remote viewing veteran and chronologer Paul Smith to clarify some of the finer points of this controversial chapter in the history of science and the intelligence world. We asked Smith about the CIA's agenda in having the American Institutes of Research (AIR) scrutinize the project, knowing full well the negative assessment that would result. Smith spoke to us from his home in Austin, Texas.

GONZO SCIENCE: When the remote viewing program was declassified in 1995, the American Institutes of Research made a report. Remote viewing insider Skip Atwater has said something about the

CIA facilitating the AIR getting this information, knowing full well that they would pooh-pooh it. Is that true?

SMITH: Well I think that's probably true; the CIA hired AIR. Now, that relationship goes back a bit further, or at least the connection with the AIR and skeptical reviews of the remote viewing program does. Some of the folks that were associated with the American Institutes of Research were also part of an earlier task force put together by the National Research Council (NRC) beginning in 1984 to review the remote viewing program at that time—and a lot of other "leading edge" human technology that the Army had been looking into besides. . . . The NRC published two books reporting their findings, which of course were completely negative regarding parapsychology. Unfortunately, there was evidence that the NRC had suppressed data that was favorable to psi while emphasizing less rigorous material that happened to be negative. There was a big stink about it at the time.

 The person who was the most important in that first task force was Ray Hyman. And he was again brought in to the AIR study by one of the leading figures at AIR who had also been some sort of coordinator on the previous NRC report. So there's a little bit of incest going on there, as you can tell. Some of these people had a longstanding track record of downplaying parapsychology, being skeptical of it, and writing against it.

 In fact, Ray Hyman goes back as far as 1972, when he and one of the people who was also involved in the later report came to the SRI labs at the time the SRI scientists were studying Uri Geller. They asked to be allowed to observe the research that was going on with Geller, but that request was declined, for several good reasons. At that time, [Hal] Puthoff and [Russel] Targ were in charge; they wanted to keep things "clean." They were getting requests on almost a daily basis for people to come in and observe the research. They were a bit suspicious of Geller at the time, and weren't sure whether what he was doing was legitimate or not, and so they didn't want to add further complicators that would make it easier for him to pull any tricks on them. Having new visitors every day adds a certain level of confusion, which would perhaps create opportunities for subterfuge or other "sleights-of-hand." It wasn't that they were convinced that Geller was trying to fool them, but they wanted to protect against that in case it turned out that he was. That's a very legitimate scientific reason.

Well, Hyman was annoyed by that. Targ told him, "We don't want to let you into our lab, but we will be happy to make Geller available to you and you can do some of your own experiments." So Hyman and his colleagues went ahead and did a couple of their own impromptu, ad hoc experiments with Geller and that was the extent of it. Later on, Hyman provided a letter he had written to the science editor at *Time* magazine and said that the work at SRI was incredibly sloppy and not well done and made a number of allegations. Of course, he had never actually been privy to the work he was criticizing.

GONZO SCIENCE: The typical skeptical sleight-of-hand, right?

SMITH: I'd say, yes. Anyway, that gives you a little bit of background on that aspect of it.

GONZO SCIENCE: Let us just make sure we have our facts straight. So the AIR is the American Institutes of Research. Would you classify them as a skeptical organization or as a civilian research association with ties to the skeptics, or what?

SMITH: They are a private organization, but they are government affiliated in a certain way in that they get a lot of government contracts. They do other research; they weren't put together to do just this review of remote viewing.

GONZO SCIENCE: So the official starting date for remote viewing research was in the early 1970s?

SMITH: The very first experiment was done in June of 1972. The program was funded by the CIA starting in the fall of that same year.

GONZO SCIENCE: So would you say that this was motivated by reports that the Russians were up to this kind of thing?

SMITH: Yes, in fact, very extensive reports. Those reports—some of them—have been declassified and are available on the Defense Intelligence Agency website.

GONZO SCIENCE: We're still curious about to what degree we can say that the CIA brought in the AIR. Did they know that they would be skeptical, and if so, how did that serve their interest?

SMITH: Here's how that worked. In the 1994 budget markup from Congress, language was inserted that would pass the remote viewing program responsibility from DIA [Defense Intelligence Agency] to CIA. DIA wanted to get rid of it. At the time these political machinations began, the commander of DIA was the general who had kicked RV out of the Army in the first place, when he'd been commander of INSCOM [Intelligence Security Command]. So when the budget was finally passed, the CIA was obligated by act of Congress to accept the RV program—then called the STAR GATE Program—but the Agency did not want to keep it. [John] Deutch—the director of the CIA at that time—was very much biased against the program, and there were a number of people in the hierarchy who were as well. So they basically told Congress—and obviously this is paraphrased—"Well, you want us to accept this program, but we want to have the opportunity to evaluate it first." At that point the CIA contracted with the AIR to do the review, and I think it cost somewhere in the neighborhood of half a million dollars, although I'm not sure of that figure.

Now, whatever passed quietly behind the scenes between the two of them is unknown, since no one in the CIA or the AIR is talking. But the internal evidence of the report itself, as well as testimony from Ed May and Jessica Utts, both of whom were involved in the review in one way or another—Utts was a significant contributor; Ed May was called on as a consultant because he had the data—those two folks both assert that there were intentional things done to limit the amount of data that would be reviewed, and limit the degree of objectivity that would be shown it. Ed May has actually written about that in an article in the *Journal of Scientific Exploration*. May talks about his involvement in the AIR study and some of the handicaps that were placed upon him and on others in the process of trying to allegedly evaluate the remote viewing program.

GONZO SCIENCE: So the skeptics in the CIA wanted a negative report because they weren't interested in remote viewing?

SMITH: It was kind of a lose-win situation for the CIA. If they were forced to actually do remote viewing they would have looked at that as a loss because, for one thing, they had abandoned the remote viewing program the first time in 1975. That was when the Church Committee was meeting and they were finding out all these bizarre things the CIA had been up to, such as the MK-ULTRA [mind control] stuff, and so on. At that point in 1975 the director of the CIA said, "I want you to get out of everything that is either illegal or controversial." Obviously, remote viewing was controversial, so they dumped it.

GONZO SCIENCE: Is that when it passed to the DIA?

SMITH: No. Actually, the Air Force, specifically FTD [Foreign Technology Division] at Wright-Patterson Air Force base acted as the funders for the program for a number of years. DIA picked it up when the Air Force jumped out of it somewhere around 1979. The DIA had been peripherally involved starting about 1975, but the Air Force was the primary there.

GONZO SCIENCE: So it kind of got passed around like a hot potato. When it changed hands, is it safe to assume that personnel within those agencies kept doing it?

SMITH: Actually, the only time that happened was when the Army had it, and then decided to get rid of it and it was passed to the DIA;

at that point the personnel stayed the same. When the Air Force got out of it, the Chief of Staff for the Air Force in fact forbade any Air Force people to be involved any further. So anyone who had been involved could no longer have anything to do with RV. That was different for the research folks out in California. They of course stayed the same. But as far as military personnel, the Army started all over with fresh people. But they continued on after that, even during the switch from Army to DIA—not counting routine transfers and such, of course. At any rate, in 1994-1995, at the time when the CIA was then being mandated by Congress to take over the program, the Agency had been caught with its hand in the pickle jar again—I don't remember what all the details were, I think it had to do with old Iran-contra stuff, and other major revelations, and they had some part in the assassination of the spouse of an American citizen; I don't remember the details of that either—but the Agency was under scrutiny again by Congress. Then right in the middle of this controversy, here comes that hot potato again, the remote viewing program. [Robert] Gates, who had preceded Deutch as DCI—with Woolsey in between them—and then Deutch himself, both had personal objections to remote viewing. And so they were apparently wondering "How do we get rid of it?" One way was to fund a study that would show that RV was of no use. Now, the "win" part of it is that if they could show it was of no use, they could still accept the program, cease doing activity, but they got to keep the personnel spaces that came along with it. So in other words, at a time when the intelligence establishment was drawing down, this was like somebody giving them a present of 10 or 12 spaces that they could use anywhere they want, if they can show satisfactorily that remote viewing was of no use, that the actual operational part of the program was of no use. So STAR GATE [the remote viewing codename] technically probably still exists in the CIA infrastructure, but only as a placeholder against which to assign these personnel spaces.

GONZO SCIENCE: How bizarre.

SMITH: But very clever from a bureaucratic standpoint.

GONZO SCIENCE: So tell us, were you ever personally involved with the program?

SMITH: Yes. I retired from the Army in 1996. I was assigned as a remote viewer from 1983 to 1990 in the program. In fact, I only left there in a big hurry because of Desert Storm; they sent me off to the 101st Airborne. . . . Let me for a moment go back to the AIR, because there's one salient fact here that's important for people to realize. The folks who did the AIR report never looked at any of the archival data from almost 20 years of operational remote viewing. They didn't look at any of that stuff when they came to their conclusion that remote viewing was of no use operationally. In other words, they did not look at the very data that would have told them one way or another whether remote viewing really had been useful. Yet even without looking at the data they declared that it had never been useful as an intelligence tool. Go figure!

GONZO SCIENCE: But they had access to it?

SMITH: That's the question. I kind of think maybe they didn't, because it was still classified and as far as I know nobody on that committee was cleared for anything. Which tells you again that the CIA knew what they were doing, that they intentionally wanted the conclusion to be negative. Because to do a thorough review they would have needed to have people cleared to do it, and give them access to all that data. It's very conveniently set up so that you can't get the hard data, except what people who were in the program verbally report, and of course that becomes anecdotal and by current scientific standards of little use as evidence. Jim Schnabel's book [*Remote Viewers: The Secret History of America's Psychic Spies*] has a number of reports of operations that are pretty close to accurate. In fact, he got some stuff out that I had no clue was going to get out; I was really impressed at what he managed to find. There will be more in the relatively near future that comes out about operations. There is some talk that finally the documents, about 120,000 pages worth, will be declassified. In fact, word is that they actually have been declassified; they just haven't been released yet by the CIA. And my guess is that some of the stuff that is happening now is going to delay that release, unfortunately. I will be publishing a book which will include a bunch of data that haven't been available thus far, but at the moment I don't want to talk about it, which I think you'll understand. But you almost don't even need to show that to show the insincerity of the CIA in this situation, because you can make a case that they did not release any of the operational data to the reviewers—or if they did, the reviewers refused to look at it—which is what would have been necessary to make an informed judgment. So the final judgment on that report was based on incomplete data, and, in fact, not even very useful data.

Now, the argument the AIR folks will make is to say, "Well, we did have operational data; we got five projects in the 12 months we were doing it, from five operational intelligence organizations that wanted work done. And we had the STAR GATE folks do these, and they uniformly came up without any useful information." That would be their argument. Well, of course, there are a couple ways of refuting that claim. First thing is, yeah, five projects—out of a total of 300 projects that the unit worked over its history! What kind of a measure is that? Second of all, at the time they tasked the STAR GATE unit, virtually all the remote viewers were gone. There were three left, and two of them were ineffective as viewers for the most part. But the final thing is, it turns out, one of the taskings—and it's in the AIR's own, unclassified report, so I can talk about this—one of the taskings was to describe and locate North Korean tunnels under the DMZ. The conclusion? The targets were worked, the results were then evaluated against known data at the time, and the viewers did describe—and remember the viewers were working blind on this, they have no clue what the target is—nonetheless, in the report, the intelligence customer admitted that the viewers did describe equipment that very closely resembled tunnel-boring equipment that North Koreans used. So that's one thing. It's not new information, but given the fact that the viewers are working blind, that is important information as far as the usefulness of the product. The second thing was that—like I said, they evaluated the tunnel information against what was known at the time.

Well, one of those viewers continues to be at DIA. And a year after everything was cut and dried and the RV unit was gone, DIA found new data on North Korean tunnels they did not have before, and it turned out to correlate very closely with what that viewer had reported. So in fact, at least one of those five operational projects produced legitimate data unknown to the intelligence community at the time. Of course it's too late for the report and all that.

GONZO SCIENCE: We know there were experiments remote viewing into and from Faraday cages, and it still worked. Was there any augmentation with different types of technology? And a correlate to that question would be. . .

SMITH: Drugs.

GONZO SCIENCE: Well, drugs. . .

SMITH: This question comes up a lot.

GONZO SCIENCE: But also, it seemed like there were remote viewing personnel who were associated with electromagnetic technology for brainwave entrancement, things like that. There are people we think who are of the opinion that, even if it's all a big fraud it went on for so long and so much money was spent that they must have been up to something else. Like maybe it was a front for some of these non-lethal weapons programs, or…

SMITH: That was never associated with the remote viewing unit, or with the research. I can say that with complete confidence that they didn't get involved in that. Now, they did of course do EEG types of studies, and biofeedback kinds of things, to see if they could see what the mechanism was that allowed for this information transfer. All the way across the board, all those studies were inconclusive. They couldn't identify any kind of mental correlates using the monitoring equipment they had at their disposal to indicate what was going on and how it was happening.…

GONZO SCIENCE: What is your take on some of the things like encountering discarnate entities and that sort of thing? The spiritual application of RV.

SMITH: I actually have a paper on my website which covers my take on the spiritual applications of it. I'm a Mormon, and my paper is called "Confessions of a Mormon Psychic Spy." But to go back to the more specific question about discarnate entities—I don't rule it out as a possibility; I've never met one. I keep an open mind, but a little bit of a skeptical mind, about the reports of remote viewers meeting so-called discarnate entities. I don't say that they didn't meet them, but I also know how remote viewing works, and it's quite possible for people to fool themselves about that. So I'm "agnostic," I guess is a good thing to say here.

GONZO SCIENCE: Contrary to what the skeptics would have us believe, it is possible to quantify, or to make a quantitative analysis of, this sort of thing. The skeptics toe this line. It's the same reason they would never deign to take a psychedelic drug—they would compromise their objectivity.

SMITH: They're looking for quantification, and of course quantification is this Holy Grail in science. "If you can't quantify it, then you always have to take it with skepticism"—which is actually a silly thing to say, because there are actually a lot of qualitative analyses done in science. And it's accepted in those areas where quantitative analysis is less useful or not feasible. I think we've fallen in love with this notion that mathematics is the only pure, *a priori* thing that there is, it's the only thing we can rely on completely and, therefore, the closer you can get to pure mathematics, the more likely you are to be able to trust the data. Again, that's baloney! There are times when qualitative analysis is not only your only tool, but it's your best tool. The example I like to use is sexing chickens. Of course there is an objective fact to be known there: you can, in a laboratory analysis, determine if one is a male or is a female. But to the casual or even the less-than-casual observer, it's very hard to tell the difference between one chick and another. Yet there are these people who seem to be able to do it. They cannot tell anybody how they do it, they just do it. Now I don't think it's anything psychic, I think it's just a really subtle, intuitional kind of thing. But nonetheless, there are things that just aren't quantifiable, and that's one of them. There are plenty of arts and skills in human society, and there are things in science. For example, morphology, where you study the forms and structures of plants or animals: it's not something you can really quantify in statistical terms. It's something you can only go by observation and comparison, yet there are no issues with that; that's perfectly acceptable from a scientific perspective.

Of course, there are obvious cautions you have to exercise. In comparing remote viewing results, you can come to convince yourself if you're not careful that the data provided in a bad session is somehow descriptive of the target. This is one form of what's called "selective validation." But such problems are well known, and the issues associated with them are well known, and it is quite possible to use similar comparative-analysis kinds of approaches that are used in morphological studies to evaluate remote viewing results in a way that may not be quantitative, but nonetheless is still a legitimate evaluation of the RV data, without falling into the evaluation traps that skeptics are fond of appealing to when trying to fault evidence for remote viewing.

GONZO SCIENCE: Do the government or the intelligence agencies have an official position on whether this was a big waste of time?

SMITH: The CIA has an official announcement that is available on their web page. It's only about a paragraph long and it says the CIA used it in the early 1970s, and evaluated it, and found it not to be effective. That's what they say. No other agency has made any official statement that I know of—if you ask, they just reference the CIA's statement.

GONZO SCIENCE: So the official party line is that, "We did some work, and it didn't work out."

SMITH: Right.

GONZO SCIENCE: So, remote viewers never tried to augment their abilities with psychedelics or entheogens or anything like that?

SMITH: Not with drugs. Now, the Soviets purportedly tried that, and they apparently managed to kill a bunch of folks, or make them sick or something, I don't know what.

I've asked Hal Puthoff this question, when I've had him lecture to people to whom I'm teaching remote viewing. When I do a basic course, I engage Hal to give a lecture to my students, since he lives here in Austin, too. He lectures on the early days of remote viewing, and in one of these lectures, I actually brought up the subject of psychoactive drugs because it had been asked so many times. And he said—and this makes eminent sense—that the whole goal of remote viewing is to be able to sort out the true signal from all the mental noise that's going on—images, internal dialogue, and stuff that all of us live with that tends to obscure the remote viewing signal. And the object is to get the signal and separate the signal from the noise, right? So the problem is that most of these psychoactive drugs actually foster hallucinations, foster a lot of mental activity, sometimes in a very unpredictable way, which only serves to increase the noise ratio. That means that drugs actually tend to make remote viewing worse, not better.

PART 9:
Parapolitics

CHAPTER 35

The Science-Swastika Connection

WHAT IS THE APPROPRIATE RELATIONSHIP BETWEEN science and the military? This piece traces the ominous overlapping gray zones between science and the military: from modern bio-electric experiments to fringe science; from the U.S. Navy to a mysterious expatriated Nazi scientist.

The "Retirement" of Dietrich Beischer

Robert O. Becker is a hero to gonzo scientists everywhere. An orthopedic surgeon twice nominated for the Nobel Prize, he is unquestionably a leader in the field of biological electricity and regeneration. His work breathed new life into ideas discarded by science in the eighteenth century, namely that electricity is vital to the life processes. The further his career progressed, the more he undermined the mechanistic concept of the body. His story is punctuated by many illuminating episodes of his struggle against heavy-handed scientific bureaucracy at its worst.

In the early 1960s, Becker began to make headway in demonstrating that nervous systems function on semiconducting DC current. Much of his work revolved around his favorite animal and test subject, the salamander. By conclusively determining that the so-called "current of injury" (present when a salamander regenerates a lost limb) was no mere side effect, Becker made great strides in understanding the basis of all regeneration and healing (as detailed elsewhere in this book).

Becker solidified his hypothesis by using DC current to anesthetize and reawaken salamanders. Following this line of thinking and trying to shore up the quality of his data, Becker proceeded to anesthetize a salamander using a magnetic field.

The concept of biological effects from magnetic fields was a radical one at that time. Becker therefore described himself as "flabbergasted" when he received an invitation from a prominent biologist to present his research at an international conference at MIT.

This conference on high-energy magnetic fields had yielded to the insistence, on the part of many reputable scientists from different countries, that a session be included on biological effects. Becker, charged with finding other presenters, received a call from a man with a thick German accent who introduced himself as Dietrich Beischer.

Beischer explained that he was familiar with Becker's work and that they had common interests. He went on to say that he was working for the Navy studying magnetic bio effects, and that much of the research was not openly published. Becker visited Beischer's lab, which was housed in the compass calibration building of the Naval Surface Weapons Center in Silver Spring, Maryland. Becker recalls being impressed with the massive resources at Beischer's disposal in the ambitious experimental attempt to null out the Earth's magnetic field and study the effects on volunteers living in the null zone.

Ten years later, Becker was serving as an outside expert on a committee reviewing a Navy report on a gigantic proposal called Project SANGUINE: to lay down 6,000 miles of cable which would comprise an enormous antenna for communicating with submerged submarines. The project was slated to cover nearly 41 percent of the state of Wisconsin, or 22,500 miles, as well as the northern half of Michigan.

By now, the American public was increasingly concerned about health effects relating to electromagnetic fields, and Becker's committee was charged with assessing the potential risk to people, livestock, and crops from the electromagnetic frequencies involved.

Becker and his colleagues cited one particularly ominous set of results in their recommendation about Project SANGUINE, later called SEAFARER. It seems that Dietrich Beischer, now operating out of Pensacola, Florida, had found that exposures of one day to the magnetic field of the SANGUINE system caused a substantial rise in serum triglyceride levels.

This qualified as a biological hazard, since serum triglyceride levels elevate in response to biological stresses, and are associated with the metabolism of fat and cholesterol. Clearly, a detrimental health effect from the SANGUINE system had been stumbled upon. The Navy took this seriously enough to run tests on the workers at their test antenna, all of whom showed elevated levels of serum triglyceride.

In a meeting with the Navy, Becker's committee unanimously recommended that the Electromagnetic Radiation Management Advisory Council be advised of the findings and the potential risk to the American public. They also recommended follow-up studies.

Incredibly, the Navy denied that that meeting had ever occurred. It also adamantly claimed that it had no information about any research indicating possible harmful effects from SANGUINE. The project wound up being shelved indefinitely, replaced by the less intrusive—but still highly contentious—ELF system in use today.

This peculiar parable is our entry point into one of the darkest chapters in the history of modern science.

From Becker's *The Body Electric*: "I'm at a pay phone. I can't talk long. They are watching me. I can't come to the meeting or ever communicate with you again. I'm sorry. You've been a good friend. Goodbye."

So went the final conversation between Dietrich Beischer and Robert Becker. Becker received the call from Beischer before a small private conference in 1977 that both men had planned to attend. Becker further recounts calling Beischer's office in Pensacola only to be told, "'I'm sorry, there is no one here by that name,' just as in the movies. A guy who had done important research there for decades just disappeared."

Dietrich Beischer disappeared four years after the report on Project SANGUINE. Not only did Beischer disappear, but in military circles, reference to his work disappeared as well. Some Navy research from 1973 references his research on serum triglycerides, but a 1980 review covering 1968 forward has no mention of it. The author of the review, James Grisset, totally omits any mention of Beischer's 19 years of work for the Navy, despite the fact that he had co-authored the 1973 findings about SANGUINE.

Beischer was tracked down in 1984 by a British TV documentary team. He agreed to appear on camera only if no sound was recorded.

Ceril Smith and Simon Best remarked upon his comments in the book *Electromagnetic Man*: "He refused to disown his discovery that low-level magnetic fields could adversely affect the heart. He also revealed that a secret survey of workers at the Wisconsin transmitter had produced similar results. But these had been classified and never released. As for his sudden retirement, that had been for personal reasons, and any other suggestion was a gross defamation of his former employers."

The final piece of the puzzle: Beischer was one of 1,400 Nazi scientists brought to the United States after World War II as part of a covert operation known as Project Paperclip.

We consider it highly likely that Beischer was forcibly retired by his intelligence-world handlers. They did so because he presented findings contrary to the goals of the nation that had secretly grafted the Nazi scientific establishment onto itself. By erasing the data and its author, the Navy removed an impediment to Project SANGUINE, which simultaneously removed a link to the taint of Nazi science.

Project Paperclip

Project Paperclip deserves close scrutiny. That many of the Nazi fugitives who escaped through "the ratline" were scientists speaks volumes, both about the role of science in industrial society and about the ability of history to go underground and become secret.

In the wake of D-Day, waves of intelligence officers swept through Europe to seize German scientists, technical personnel, and equipment. In December 1944, Office of the Secret Service head Bill Donovan and the OSS head of intelligence in Europe, Allen Dulles, urged FDR to permit Nazi industrialists, scientists, and intelligence officers to enter the United States.

FDR responded, "We expect that the number of Germans who are anxious to save their skins and property will rapidly increase. Among them may be some who should be tried for war crimes or at least be arrested for active participation in Nazi activities. Even with the necessary controls you mention, I am not prepared to authorize the giving of guarantees."

His presidential veto was meaningless. By July 1945, a Joint Chiefs of Staff-approved plan was operating to bring 350 German scientists to the United States. Werner von Braun and his V2 rocket team were among this flow of German submarine and artillery engineers and chemical-weapons designers.

Von Braun had used slave labor at his Mittelwerk complex. More than 20,000 people had died from overwork and exhaustion. And according to an article called "Germans at Last Learn Truth About von Braun's 'Space Research Base'" that was published in the U.K.'s *Telegraph*, "A Polish slave-labour survivor of the Dora factory [where von Braun's laborers were brought from] recalls how Wernher von Braun visited the works and seemed 'completely unperturbed' by the piles of corpses."

George Rickhey, supervisor at the Dora concentration camp, was known for hanging laborers 12 at a time for acts of sabotage. Rickhey had children clubbed to death because they were useless to the rocket team. This record was no impediment to a transfer to Wright-Patterson Air Base, where he oversaw dozens of other Nazis pursuing research in the United States.

This also put him in a good position to eliminate documents that might have compromised the team. Columnist Drew Pearson helped secure public disquiet to the point where a pro-forma trial for Rickhey was necessary. This secret trial was supervised by the Army, which had every reason to suppress the fact that the Mittelwerk team of war criminals was now on the payroll. Preventing von Braun's and others' interrogation, as well as withholding documents now in the United States, the trial was sabotaged, and Rickhey was acquitted.

From Christopher Simpson's *Blowback:*

By the end of 1947 the U.S. Army had begun at least a half dozen large-scale programs designed to tap the talents of the SS and German military intelligence veterans. Operation Pajamas, for example, organized "exploitation of German personnel used in forecasting European political trends." Birchwood did the same with "economic experts," in this context clearly suggesting men who had worked for the SS and for [the infamous Hermann] Goering. Project Dwindle collected Nazi cryptographic experts and equipment. Apple Pie, a joint U.S.-British operation, recruited "certain key personnel of [SS] RSHA Amt VI" who were expert in Soviet industrial and economic matters, according to the U.S. orders that established the code word designators for the program. Project Panhandle undertook " operational exploitation"—in other words, recruitment for pay—"of German ex-Military Intelligence personnel for collecting military intelligence on the USSR and its satellites." Project Credulity traced German scientists wanted for the JIOA Paperclip project. These efforts, though highly secret from the general public, were nevertheless approved and managed through regular intelligence channels. They received conventional code names and were financed in the normal army intelligence budget. These were not a conspiracy within the intelligence community to defy the rest of the government; these exploitation programs were the official, though secret, U.S. policy (p.73).

Forces within the American government were divided between those who wanted to prosecute every Nazi and neutralize German aggression forever, and those who felt that a reintegration of Germany into the economic system as soon as possible was the desirable course.

A pressing issue at hand was identifying the offensive capability of the Soviets. Expressing skepticism about Soviet military designs on Europe was a liability. Reinhard Gehlen, a former German army general, was authorized to construct a new covert organization comprised of German experts on the USSR. This network became a primary source of Western intelligence on the Soviets. Not only did they employ former SS and collaborators, but they constantly misrepresented the Soviets' battle readiness. By purposefully overestimating the Soviet threat, they made themselves indispensable to countering that threat, and thereby created job security at the expense of exacerbating the cold war. We used them and got used in return.

It is important to recognize that the U.S. Army contributed to more prosecutions of war criminals than any other institution in the world, rivaled only by the Soviet secret police. The paradoxes at work in the vacuum of the Reich's collapse must have been immense. The relationship between the West and Russia was chilling fast. Everywhere, intelligence agents were arresting Nazis for crimes against humanity on the one hand, and exploiting Nazis for use against the Soviet threat on the other.

Intelligence agents were breaking up various webs of the Nazi underground movement or, conversely, co-opting them. The U.S. forces that tracked and prosecuted the fugitives were also responsible for certain select ones escaping justice and being put to work for America's interests. Truman had signed off on the idea of making use of certain personnel, but left the question of distinguishing "ordinary Nazis" from "war criminals" in the hands of the State Department.

Employment of Nazi personnel was almost universally justified by the now eyeball-to-eyeball relationship with the Soviet Union. While the Soviets were not above overlooking someone's role in the war when it served their purposes, they also used American coddling of highly sought-after war criminals as fodder for their propaganda. Use of Nazis thereby increased tensions between the nuclear superpowers.

The Paperclip program for repatriating Nazis to the United States targeted Kurt Blome, Hermann Becker Fryseng, Konrad Schaeffer, and Dr. Sigmund Rascher, all among the vilest purveyors of torture in the guise of science. Emil Salmon was sheltered at Wright-Patterson in Ohio after being convicted of crimes in a denazification court in Germany. There was intense competition among different American military branches and intelligence agencies for control of valuable personnel assets from the Reich.

The Americans took great interest in Dr. Kurt Plotner's use of mescaline as a method of gaining insight into a captive's psychology. Boris Pash of the newly born CIA pored over the notes from the Dachau concentration camp looking for leads on extracting information, psychosurgery, speech-inducing drugs, electroshock, etc.

Here then is the bridge: the CIA's experiments with LSD were conceived of as continuations of the research at Dachau. More than 7,000 U.S. soldiers were unwitting subjects to experimentation with psychedelic chemicals, many manifesting psychological crises and dozens attempting suicide.

America's history of experimentation with nuclear materials on its own citizens is another example.

In one notorious case involving the use of radiation to sterilize unwitting prisoners, a Dr. Joseph Hamilton candidly remarked that the experiment "had a little of the Buchenwald touch."

Through the secret policy of snapping up Nazi scientists, not only was the cold war exacerbated, but the practice of science itself was corrupted. In the words of Neal Acherson, "Those who stole the plague germs infected themselves."

Recommended Reading: *Blowback: America's Recruitment of Nazis and its Effects on the Cold War* by Christopher Simpson; *The Body Electric* by Robert O. Becker and Gary Selden; *Cross Currents* by Robert O. Becker; *Whiteout* by Alexander Cockburn and Jeffrey St. Clair

CHAPTER 36

The JFK Assassination

DEALEY PLAZA, 1963—DOES ANYONE REALLY KNOW what happened? Depending on the book one reads, a compelling case can be made for a thousand conclusions. The colossal irony is that while it is the most studied single day in the entire history of the world (quickly being overtaken by 9/11) it's still damnably hard to figure out whodunit (unless you swallow the Warren Commission, that is).

The difficulty surrounding the JFK assassination lies in the simple fact of its complexity. A maddening quantum-style uncertainty surrounds the plaza, and as the limo rounds the corner onto Elm Street, John F. Kennedy becomes focused in the crosshairs of an indeterminate number of riflescopes.

But has the case therefore devolved from a pursuit of truth into a mere intellectual puzzle, with no more meaning than a game of Tetris? The footage of that day has rolled before our eyes so many times; has the assassination become a mere pop-culture object? (The gross Zapruder film is surely either the world's greatest snuff film or the world's greatest Rorschach test.)

Conclusions shift tectonically with each assassination book's perspective. Do the trajectories of Dealey Plaza trace an undecipherable hieroglyph?

It seems pretty clear to those with even a passing familiarity with the case that Oswald did not act alone, if at all. There is some evidence that Oswald may have thought he was working *for* the U.S. intelligence machine, when in fact he was unknowingly being manipulated by unscrupulous agents of it. Author George Michael Evica states that his research indicates that Oswald actually thought he was penetrating an assassination plot as a double agent, all the while being maneuvered into place as the fall guy.

This theory, as well as its many variants, makes Oswald one of the three great victims of that day, next to JFK and Jackie.

And let's not forget the multiple wounds of Governor John Connally, who swore until his dying day—as did his wife—that he was shot at by more than one person, flatly contradicting the Warren Report.

There is a school of thought however that links Connally to the plot, if only tangentially, through LBJ. Connally and LBJ went back quite far along a seamy Dallas-Washington, D.C., axis of power, an axis that included oilmen H.L. Hunt and Clint Murchison, FBI Chief J. Edgar Hoover, and Richard Nixon, among others, who had motives for desiring an early end to the JFK presidency.

It is possible at least to make educated guesses. JFK was hated in some circles—that's no secret. The Bay of Pigs veterans in particular had means, motives, and opportunity. And only in the Bay of Pigs veterans do you see a constellation of three major groups that have separately been targeted with blame for the assassination: the CIA, the Cuban exiles, and the Mafia. We wouldn't have wanted to cross them all at one time in a single stroke, like JFK did at the Bay of Pigs. Talk about living dangerously; this guy made himself some *enemies*.

Looking under a few rocks may illuminate matters somewhat.

The CIA-JFK Connection

It turns out to be very tempting to believe that the CIA did do it after all, when one considers just how outrageous, dirty, and stupid the CIA has been during the past few decades. To inform our discussion, and to see what the CIA is capable of, we here present the JFK assassination considered as a CIA operation. It is instructive to acknowledge that the CIA has often run operations that have directly conflicted with the United States government and its executive branch. These include:

Costa Rica, 1955. The U.S. State Department let Costa Rica have fighter planes to defend itself against rebels. But who was flying the attack missions for the rebels? CIA pilots in CIA planes.

Cuba, 1957–1958. The U.S. government shipped weapons to Cuban President Batista so that he could quash a rebellion. And who was giving money to the rebels? The CIA. (The CIA lived to regret this one. The rebel was Castro, and after the CIA helped him win, he turned commie, and the CIA spent years sneaking Mafia hit men into Cuba to kill him, not to mention the whole Bay of Pigs thing. As a member of the Warren Commission, Allen Dulles made sure the other members were steered very far away from this arena of violence and anti-JFK sentiment.)

Burma, 1970. The U.S. military worked hand in hand with the Burmese government to crush rebels, who were working hand in hand with the CIA to crush the Burmese government.

Angola, 1960s–1970s. Washington supported the Angolans with military aid and counter-insurgency training to suppress a rebellion, while the rebel leader was on the CIA payroll.

Other Coup Attempts

It's clear that the national interest of the United States is not always central to the CIA's plans. It is enlightening in this regard to consider the JFK assassination in its context of worldwide CIA-sponsored coup attempts:

July 11, 1963. Four months before JFK is killed, a CIA-supported military junta overthrows left-leaning Ecuadorian President Arosemana.

November 1, 1963. Three weeks before JFK is murdered, the president of South Vietnam, Ngo Dinh Diem, is murdered in a coup planned from within the U.S. State Department. (The State Department was often a cover for CIA agents; for years it was actually run by John Foster Dulles, brother of CIA director Allen Dulles.) Diem had been installed by the CIA, but he became a liability so they knocked him off.

November 20, 1963. Two days before JFK dies, the Cambodian leader, Prince Sihanouk, initiates a vote in the Cambodian National Congress to end all U.S. aid to his country, and the CIA meddling which piggybacks on top of it.

Unprecedented in world politics, the vote is sandwiched among various dirty CIA incidents in Cambodia, including a 1963 car-bomb assassination attempt. Apparently, JFK assured Sihanouk that he had nothing to do with it, and Sihanouk believed him. It was clear to both men at this time that the CIA was a rogue elephant.

Laos, 1961-1964. Since 1961, JFK had made a sustained diplomatic effort to lash together a coalition government in Laos between left- and right-wing factions. JFK's efforts ran directly against the actions of the previous Eisenhower-Nixon administration, which had set up an ongoing CIA operation to violently crush the left. The CIA ignored JFK's wishes and kept the secret war going. Five months after the JFK assassination, the shaky coalition government gave way to the CIA's handpicked right-wing guys (and nine months after the assassination, the CIA faked the Gulf of Tonkin attack, which gave Johnson an excuse to do what JFK wouldn't: start the Vietnam War).

Bay of Pigs, April 1961. JFK's inauguration had no effect on another CIA plan-in-progress from the Eisenhower-Nixon administration, namely, the plan to spark an anti-Castro revolution by invading Cuba at the Bay of Pigs in mid-April of 1961. (This appears roughly synchronized with the April 22 CIA-backed plot to topple Algeria and De Gaulle.)

This crackbrained plan, involving a CIA-Cuban exile-Mafia-mercenary coalition, was Vice President Nixon's baby, and you can be sure he was chagrined to lose the election to JFK right before it came off. JFK, feeling manipulated by the CIA, refused to have anything to do with the invasion, and this caused the plan to fail more miserably than it would have otherwise, which caused the veterans of it to curse JFK's name with murderous rage.

Thus we see that the JFK assassination is naturally understood in the context of similar, violent CIA operations worldwide. The CIA's fingerprints are patently all over the assassination.

However, as Peter Dale Scott reminds us in *Deep Politics and the Death of JFK*: "I would not suggest that the CIA had a motive to kill Kennedy. It is a simplistic anthropomorphism to treat the CIA as if it had a single-minded point of view, when in fact the CIA is…a cockpit for conflicting viewpoints."

Historian John Newman elaborates this point in *Oswald and the CIA*, and offers the caveat that the CIA certainly had the good sense to not become *officially* involved in the plot: "We have yet to find documentary evidence for an institutional plot in the CIA to murder the president. The facts do not compel such a conclusion." But Newman does go on to say: "The facts may well fit into other scenarios, such as the 'renegade faction' hypothesis."

Some level of CIA involvement seems impossible to ignore. However, as Scott understands, the assassination was simply too big a job for one agency, or a single renegade faction, to have accomplished: "not even the CIA could have 'acted alone.' Officials of other agencies appear to have acted conspiratorially on or before November 22: the FBI…; the Secret Service;…and army intelligence (among others)."

In other words, a somewhat wider lens may be needed.

A Wider Lens: The Dallas-Watergate-FBI Connection

From *High Treason* by Robert Groden and Harrison Livingstone:

> We can begin to understand how [the JFK] assassination could have happened, how so many witnesses could have died, if we try to look at some hard and ugly facts…Ask yourself, how did Martin Luther King, Jr. die? And how did his brother die in his swimming pool? And what kind of shots or attack were made on other members of his family? What about the shooting of George Wallace in 1972, without which Richard Nixon most probably never would have been re-elected? The murder of Robert Kennedy? The murder of Allard Lowenstein who found a conspiracy in the murder of his friend, Robert Kennedy? The near murder of Edward Kennedy? …Witnesses died in the Watergate affair, in the shooting of Wallace, in the assassinations of Martin Luther King, Jr., Robert Kennedy and

John Kennedy…It has been said that all the political assassinations of the 1963–1973 period are connected, and that the liberal/centrist leadership of the country is being exterminated, as in banana republics…Clearly, the Kennedy, King and Wallace shootings were conspiracies, each supported by a cover-up…Certainly the media is unwilling to admit that they were wrong in accepting without question the Warren Report [Although Dan Rather and Peter Jennings both show indications that they do not believe every word of it—Jim & Al]—How could it have happened without someone finding out the truth? To begin with, the circle of conspirators was relatively small, but they had at their disposal all the apparatus of government and the underworld (p. 407-408).

Assuming the above has some general veracity, one of the men who figures very prominently in the above hijinks—and who had "all the apparatus of government and the underworld" at his disposal—was Richard Nixon.

Nixon's connection to the JFK assassination, it seems to us, is not a trivial one, and it followed him all the way to the White House, where he continued to surround himself with some of the dirtiest, crookedest right-wing spooks and criminals of all time, who themselves had direct links to the JFK assassination.

Nixon was totally down with all sorts of people who wanted JFK dead, namely many in the CIA, but also the Mafia and the Bay of Pigs veterans (who were themselves a hodge-podge of mob guys, CIA guys, and right-wing Cuban expatriate guys).

Nixon's mob connections are widely recognized and hard to miss. They reach at least as far back as his 1946 run for the House of Representatives, partially bankrolled by brutal Los Angeles Mafioso Mickie Cohen. Nixon also pardoned Jimmy Hoffa, one of many mobsters who made threats against the Kennedys and who also bankrolled Nixon's campaigns. The Watergate tapes reveal Nixon openly discussing laundering payoff money through his Mafia connections.

It even seems that a certain prominent mobster—Jack Ruby himself—owed Nixon a favor. In Nixon's California congressman days of the 1940s, Nixon's staff used Ruby for "information services," according to a 1947 memo discussed by researcher Gary Shaw in *Cover-Up: The Governmental Conspiracy to Conceal the Facts About the Public Execution of John Kennedy*. The 1947 memo—written by a Nixon staffer who was also an FBI agent—saved Ruby from having to testify before a congressional committee in that year. Ruby then moved to Dallas, and years later, Nixon was there too, on November 22, 1963.

Also in Dallas on that day was Nixon's old CIA buddy, fellow Bay of Pigs man, and future Nixon White House crony/Watergate burglar E. Howard Hunt. Each man initially denied being in Dallas on that day, and Hunt may still deny it, although some sworn testimony (in the U.S. District Court for the Southern District of Florida) and a CIA memo have stood against Hunt's own claims of being at home with his family. Hunt had to "remind" his own wife and daughters that he was at home. (All of this came out in a court of law as documented in Mark Lane's *Plausible Denial: Was the CIA Involved in the Assassination of JFK?*).

Nixon is known to have assigned assassination jobs to Hunt, including heads-of-state (notably, Nixon sent him to Panama to assassinate Panamanian dictator Omar Torrijos) and American citizens. Hunt and Nixon's working relationship goes at least as far back as the Eisenhower administration, when they hatched the Bay of Pigs plan to invade Cuba together.

As Eisenhower's vice president and CIA point man, Nixon was the front man for a troop of trained killers. Nixon was set to assume the presidency and invade Cuba with these guys when Kennedy stole the election. But the invasion was already in motion, and when Kennedy kiboshed it, it drove the operation's survivors to hate him. Nixon was one of these rabid men; an anti-communist who believed Kennedy was destroying the country.

Nixon might not have pulled the trigger, but it seems that Hunt could have. In all likelihood, Nixon was in on the order to kill, by virtue of being almost-president with a loyal intelligence apparatus already in place. We find it quite believable that at the very least he knew something about the assassination conspiracy before it happened.

Richard Nixon's right-hand man in the White House, H.R. Haldeman, apparently believes that Nixon and E. Howard Hunt were both connected to the JFK assassination. Haldeman wrote in his memoir that Nixon grew afraid that Hunt would start spilling secrets to save his own ass from frying when the Watergate conspiracy was exposed. Nixon paid Hunt not to talk—and shortly thereafter, Hunt's wife, with the money, and a bunch of other Watergate figures, were blown up in a distinctly dodgy airplane disaster over Chicago (which probably served as the inspiration for a similar scene in the Chuck Norris movie *Good Guys Wear Black*).

Another airplane disaster conveniently killed Warren Commissioner and "magic-bullet" theory disbeliever, Louisiana Congressman Hale Boggs, just as he was starting to publicly confront FBI head J. Edgar Hoover about the FBI wiretap of his phone.

Then Hoover conveniently died of a heart attack during a power struggle with Nixon, two weeks before the [George] Wallace shooting. The *Harvard Crimson* apparently ran a report relevant to this. As discussed in *High Treason*, this report stated that Senator Ervin's Watergate Committee received evidence that Nixon henchman G. Gordon Liddy (who was one of Nixon's "plumbers," along with Hunt) may have been Hoover's assassin, breaking into Hoover's apartment and poisoning his toiletries with heart attack-causing thyon-phosphate.

With regards to the Wallace shooting, *Emory Magazine* ran an article in its Spring 1996 issue in which it is stated that not only had Nixon sicced the IRS on Wallace to try to derail his candidacy, but after the shooting, Nixon dispatched "a former CIA operative" to the alleged shooter's apartment to try to plant McGovern campaign literature there. Just like all these other cases where the shooting cannot be directly linked to Nixon, Nixon nonetheless seems uncomfortably close to it, and is somewhat naturally caught in the web of suspicion.

Watergate was the apex of a series of political crimes that snaked back through Dallas to the Bay of Pigs. Watergate threatened to blow Nixon's secrets wide open, and his trail back to the JFK assassination would have come under public scrutiny.

Enter Gerald Ford, Nixon's obedient vice president, who had also been Hoover's obedient mole on the Warren Commission. Ford pardoned Nixon, which covered up the dirtiest secrets in American politics. The pardon covered up the full details of the complex of Watergate crimes. The pardon also helped to cover up Nixon's and Ford's lingering connections to the JFK assassination and also the corruption of the Warren Commission.

As a member of the Warren Commission, Ford had enthusiastically passed its secret deliberations to J. Edgar Hoover. If Nixon's Watergate connections lead straight back to Dealey Plaza, and if Ford helped cover them up as a member of the Warren Commission, then Ford's pardon of Nixon was self-interest at work.

The crowning irony is that the Nixon pardon won Ford the John F. Kennedy Courage Award for his "act of conscience."

Why did Ford pass the Warren Commission's secrets to Hoover? The commission had no investigative powers—it relied solely on the FBI's information. And the FBI had things to hide in this matter, and didn't want any of the Bureau's numerous assassination connections to be uncovered. The FBI, in the person of J. Edgar Hoover, wanted to know what the commission knew so it could be more easily led down very precise avenues of misinformation. Hoover needed a friend on the commission. Ford, ever eager to feel like an intelligence-world player, was his boy.

As an illustration of Ford's love affair with cloak-and-dagger intrigue, it is worthwhile to take note of his behavior as president, when he reluctantly found himself forced to investigate the excesses of the CIA and the FBI. His response was to give these out-of-control agencies whatever cover he could. According to Kathryn S. Olmsted's book *Challenging the Secret Government: The Post-Watergate Investigations of the CIA and FBI*, Ford slyly turned officially recommended intelligence community reforms on their heads. Olmstead writes:

> Ford proposed a minor reorganization of the
> intelligence community…But the thrust of his
> proposals was to *strengthen* [emphasis in original]
> the agencies and the laws protecting their secrecy.
> The president's response, in many cases, to the
> revelations of domestic spying by the CIA was to
> make that spying legal…[Prominent intelligence
> community investigator Senator Frank Church]
> commented that [Ford's changes] gave the CIA "a
> bigger shield and a longer sword with which to
> stab about" (p. 173).

The CIA wasn't the only intelligence agency with an uncomfortably close connection to the assassination. The FBI had watched/used Lee Harvey Oswald for years. Oswald's professional FBI informant status—and his working relationship with FBI agent James Hosty—would have greatly embarrassed the FBI if this information had been made public at the time of the assassination.

Three days before the assassination, Oswald contacted his FBI man, James Hosty, and gave him a note. When Oswald was killed by Ruby within a week, word came to Hosty: destroy the note. The *New York Times* ran an article saying Hoover himself gave the order to destroy the note. What haunts us the most is researcher George Michael Evica's contention that the note might very well have said something like, "I have infiltrated a group who conspire to kill JFK in Dealey Plaza. Please warn the president!"

FBI chief J. Edgar Hoover, like Nixon, didn't pull any triggers. But like Nixon, Hoover could have known about the assassination beforehand, or who did it. Both Nixon and Hoover had strong right-wing intelligence/Mafia/Cuban-expatriate ties, exactly the same milieu that generates the most connections to JFK's assassination.

Peter Dale Scott has come to much the same conclusion in *Deep Politics and the Death of JFK*:

Of all the components to this nexus of anti-Kennedy intrigues, the most crucial and powerful input, both before and after the assassination, was that of J. Edgar Hoover…a key to the successful murder of the President, and the ensuing cover-up, is the deep political counter-coalition of intrigue…involving Hoover…There is no comparable suspect…[I]n 1963, as the tensions increased between the Kennedys and their enemies, the various anti-Kennedy coalitions found themselves more and more in a common camp, with Hoover as their strongest ally (p. 224-225).

In other words, Hoover's role in planning the assassination was perhaps more central than anybody else's. This central role becomes even more haunting and plausible when one considers the parallels between the death of JFK and another one of Hoover's hated enemies, Martin Luther King, Jr.

The JFK-MLK Connection

JFK's assassination, like Martin Luther King's, was made possible by the lifting of his security. Even a cursory glance at JFK's Dallas security situation suggests that his assassination was carefully arranged with the help of some very powerful people, specifically—in this regard—members of the Secret Service.

Not only did Secret Service agent Forrest Sorrels change the motorcade route to include ambush-friendly Dealey Plaza, but in concert with this, agent Winston Lawson trimmed the motorcycle entourage from eight motorcycles to four, and each of these four motorcycles was deployed strictly to the rear of the presidential limo. These actions by agents Sorrels and Lawson steered JFK into a dangerous spot surrounded by tall buildings, and simultaneously stripped him of the shielding protection of the motorcycles, which is precisely what the motorcycles are there for.

No Secret Service agents were stationed in Dealey Plaza itself, nor did any of them give it even the most basic preliminary once-over before the motorcade. In fact, the *Fort-Worth Star Telegram* reported that at least nine but as many as 17 Secret Service agents were up until 3:30 a.m. the night before the motorcade getting absolutely bombed at a local bar. At least four of these agents had "key responsibilities" relating to the security of the president's limo the next day.

This escapade explains why—in the words of Senator Ralph Yarborough—"all of the Secret Service men seemed . . . to respond very slowly, with no more than a puzzled look." (Senator Yarborough was seated in the same car as LBJ, and says that Johnson was listening intently to a walkie-talkie—the volume down low—all during the motorcade. In this regard, in the book *The Texas Connection: The Assassination of John F. Kennedy*, researcher Craig I. Zerbel discusses the fact that LBJ and JFK had fought bitterly the night before. Johnson wanted his friend Connally to ride back in LBJ's car, and tried hard to get Yarborough—a political enemy—placed in the JFK limo in Connally's stead.)

The limo driver—agent Bill Greer—slowed the car to a crawl for nearly ten seconds right as the shooting began.

These members of the Secret Service probably didn't even know who was shooting. But it seems as if they all knew the roles they had to play in order to make a shooting successful.

In MLK's case, the "stripping of security" aspect figures prominently in its comparison to the JFK assassination. In a previous visit to Memphis, King had retreated to a secure but white-owned hotel when riots broke out during civil rights demonstrations. As he made his return to Memphis for the next round, a well-timed newspaper article appeared that publicly pressured King to avoid the appearance of hypocrisy and make minority-owned accommodations. King was thereby prodded into his less-secure arrangements. But the King party had booked rooms on the lower floor of this hotel, until an unidentified person showed up early and changed the reservations to a higher floor, with balconies totally exposed to the surrounding buildings. In addition, the area's only black police officer and black firefighter-paramedics received last-minute transfers away from the vicinity, completing King's isolation and exposure.

In both the JFK and MLK assassinations, a rifle and scope were found abandoned at the scene. The alleged riflemen are also linked through their possession of too much money. James Earl Ray was destitute and yet somehow managed to spend thousands of dollars escaping to Europe. This exactly parallels Oswald's fake defection to Russia, when he had $203 to his name and yet spent $1,500 to get overseas. Oswald continued to spend beyond his means up to the assassination, and of course it is by now well established in the historical record that Oswald was an intelligence agent for at least a couple of different agencies and acted as an informant for the FBI.

This is why Oswald's tax returns are still top secret: his extra money came from the U.S. intelligence underworld, and it was the same underworld that manipulated him into the role of patsy, and it is the same underworld whose connections to the assassination would be exposed right there on Oswald's tax returns. The parallel to James Earl Ray's mysterious extra money seems apparent, as he himself claimed to be a patsy, as Oswald did in his same position of intelligence-underworld pawn.

In each assassination, we confront the involvement, at some level, of J. Edgar Hoover. Hoover was a personal enemy of the assassinated men and was blackmailing them too—and half of Washington, D.C., as well. Hoover did his best to obstruct the Warren Commission's investigation with the help of his insider stooge, Gerald Ford. And Hoover had also already tried to kill MLK once before with an FBI plan to get King to commit suicide. It took the form of a note that threatened exposure of extramarital affairs, saying stuff like, "There is only one way out for your filthy self."

Kennedy, King, Oswald, and James Earl Ray were all carefully maneuvered into position by a constellation of powerful people from behind the scenes.

Recommended Reading: *Who Shot JFK? A Guide to the Major Conspiracy Theories* by Bob Callahan; *Plausible Denial* by Mark Lane; *Coup D'Etat in America* by Alan J. Weberman and Michael Canfield; *Target: De Gaulle* by Christian Plume and Pierre Demaret; *Interim Report: Alleged Assassination Plots Involving Foreign Leaders* by the U.S. Senate Select Committee to Study Governmental Operations with Respect to Intelligence Activities; *Killing Hope: U.S. Military and CIA Interventions Since World War II* by William Blum; *In the Midst of Wars* by Edward G. Lansdale; *Air America* by Christopher Robbins

CHAPTER 37

Interview with JFK Assassination Researcher Jim Fetzer

GONZO SCIENCE INTERVIEWED JFK ASSASSINATION researcher Jim Fetzer in his office at University of Minnesota-Duluth on October 25, 2002. The news about Senator Paul Wellstone's fatal crash had just broken a couple of hours previously. The coincidence was oppressive and terrible and we nearly put off the interview.

Fetzer's knowledge of the JFK assassination is truly encyclopedic. He has read every book, seen every scrap of film, worked closely with other experts, edited books on the topic, conducted symposiums, and interviewed assassination insiders. From all of this he has come to the conclusion that JFK was killed as the result of a massive government conspiracy. We spoke with him at length about the various levels of the conspiracy as he has come to understand it.

GONZO SCIENCE: So how many shooters were there, and how many shots?

FETZER: There appear to have been at least six shooters—six who actually shot—and there were nine, ten, or 11 shots.

Essentially, you've got the limo coming up Houston turning onto Elm. If Oswald had been located in the alleged assassin's lair, these would have been his best possible shots. You've got the president getting closer and closer to him; his chest is exposed, his head is exposed.

Now, I used to supervise recruit training; I was a Marine Corps officer. Same recruit depot, same rifle range—Edson Range, Camp Pendleton, Marine Corps Recruit Depot, San Diego—where

Oswald took his training.

GONZO SCIENCE: Really.

FETZER: Yeah, by coincidence. I know, I wouldn't think it would come into play here. But he was definitely a mediocre shot.

In 1957, he qualified with a 212, which isn't too bad. That's sharpshooter; it's the second-highest rating in the Marine Corps. It's all based on 250 points: 50 points from five different positions and distances. I shot 212 on more than one occasion. And the next year [Oswald] apparently didn't qualify at all, even though Marines are required to qualify every year with a rifle. And the next year after that he qualified with only a 191.

Now 190 is the minimum range for achieving marksmanship status. And he probably was given that, because if you're working the pits, and somebody's just a couple of points away from qualifying, they tend to give it to you. His buddies in the Corps were talking about him "getting Maggie's drawers," which is this white flag that they wave when you miss the target completely. He was a shit shot! Plus, marksmanship is a very practice-oriented activity, I mean, you've got to be practicing a lot.

This piece of weapon [Oswald's Mannlicher-Carcano rifle, the alleged JFK assassination weapon] was a piece of junk. It wasn't even high velocity.

Now here's the simplest possible exoneration of Oswald. The president was killed by the impact of high-velocity rounds. The damage to his cranium and brain was so extensive, it could not have had any other cause. But the Mannlicher-Carcano only has a muzzle velocity of 2,000 feet per second; by comparison, this Bushwacker .223 [the weapon of the Malvo-Mohammed "Washington-area sniper" duo, recently captured at the time of this interview] has a muzzle velocity of 3,000 feet per second. That's a high-velocity weapon. But the Mannlicher-Carcano is not.

So, you tell me, how can a weapon which only has a 2,000-foot-per-second muzzle velocity fire high-velocity bullets? The answer is that it cannot. And this is the only weapon that they ever tried to nail on Oswald, even though [Dallas Police] Chief Curry, two days after the assassination, was still saying, "We got no way of tying him to the weapon."

The FBI went over [the rifle] twice, couldn't find any fingerprints, and sent it back to Dallas. They took it out to the mortuary, put ink on [Oswald's] hands, and put a print on the barrel underneath the hand guard. And the guy who ran the morgue said he had a hell of a time getting that ink off.

GONZO SCIENCE: I never read that anywhere.

FETZER: This is not news, in another sense, right? So you got this limo coming up. You've got Oswald not taking the shots; they would have been irresistible shots. The driver turns the corner on Elm, but he swings up too widely to the right, and nearly hits a concrete abutment. He virtually brings the limousine to a halt right there.

GONZO SCIENCE: And this is based on eyewitness testimony.

FETZER: Oh yeah, yeah. [In addition to 60 other eyewitnesses] this is Roy Truly, Oswald's supervisor at the Depository; he was a wonderful witness to this.

And so here's Oswald…Oswald was never taught to use a scope, he wasn't taught to fire with a bolt-action rifle. He was taught to shoot with an M-1, which is semi-automatic; each time you pull the trigger, it automatically shoots. This [Mannlicher-Carcano] is a bolt-action thing, and the bolt was so stiff that when they asked master marksmen to replicate what Oswald had done, they found they simply could not do it, and when they worked the bolt, it pulled them off the target.

Now when you use a scope, it does enhance accuracy, but it decreases time—it takes more time to get those fine little hairs set, see. So when you're trying to sight in with a scope, you've got to take this time, and if you can get that fucker there, and then if you know what you're doing, right, if you have a gentle trigger squeeze—and this weapon had a funny double-action trigger squeeze, so at first it sort of was normal, then all of a sudden it jerked, then all of a sudden it would jerk again.

It also had what has been described as "the coy habit of blowing the firing pin out in the face of the shooter."…They had to put shims in; the scope was misaligned, they had to rebuild it with shims before these marksmen would even try to fire with it because it was so far out of alignment.

I mean it was a complete piece of junk, these things sold by the bushel basket, you know, ten dollars for a dozen, that kind of shit.

GONZO SCIENCE: Didn't the Italians call it the "pacifist rifle"?

FETZER: The "humanitarian rifle," for never injuring anyone on purpose.

GONZO SCIENCE: "The world's worst rifle."

FETZER: In my opinion, it was. Terrible. The Warren Commission had a lot of trouble finding anyone who would say it was a decent piece of hardware, but they eventually tracked down some gunnery sergeant and some officer, I'm embarrassed to say, who would say it was a good weapon, but everyone knew it wasn't. It was a piece of shit.

So you get Oswald—he wasn't even on the sixth floor, by the way, he was actually seen in and around the first floor and the second floor at the time of the assassination. He was actually drinking a Coke when he was apprehended there by a motorcycle patrolman who rushed up within 90 seconds of the assassination and confronted him in the lunchroom.

Now that's curious all by itself because the lunchroom was way in back; it's a rather obscure area. Why this motorcycle patrolman, who wants to secure the building because he thinks there's a sniper there, is going back to the lunchroom on the first or the second floor, is very peculiar. Why isn't he racing up the stairs to the sixth floor where he thinks the assassin was?

And here's Oswald, who's just finished his lunch—and you have three or four employees who saw him at a quarter to noon, at noon, and as late as 25 minutes after noon, in and around the lunchroom. Then he's confronted by the motorcycle patrolman within 90 seconds after the assassination.

So how does he get up and down and shoot the president? What's the first thing he does? He rushes across the warehouse floor, stashes his trusty Mannlicher-Carcano, races down four flights of stairs, for what? To have a Coke.

So I say, hey, the commercial possibilities are endless. You have Oswald with his Mannlicher-Carcano in one hand, his Coke in the other, saying, "Things really do go better!"

And they put pressure on the motorcycle patrolman who, when he wrote up his initial report, said, of course, "The guy had a Coke." They told him to take it out. So he scratched it out. . . . Pressure was put on him to scratch it out, but you can actually see this handwritten report of his where he scratched it out.

So okay, so here's the driver, he's brought his limo to a halt, Oswald isn't there anyway, I mean, these are disciplined marksmen scattered all over the place. They're waiting until [JFK] gets right in the kill zone.

On the left-hand curb actually there are three yellow stripes that are marked that I am convinced are designated areas for shots— I think probably it was intended to be three volleys. Like when you're opposite the first yellow mark—I never saw them when they were unfaded—but…you use them in relation to the limousine, either when they correspond to the yellow mark then you shoot, or when they're between the yellow marks.

And then you've got this umbrella man, pumping his umbrella up and down, indicating that you keep shooting because the target's not dead, right? And then you've got this guy, this Latino guy, shaking his fist at JFK, you know, and undoubtedly denouncing him, "*Sic semper tyrannis!*" He might as well be John Wilkes Booth, right? Except you've got these guys scattered all over.

So [JFK] starts driving down Elm Street. The first shot, they say, sounded different. The first shot then had to be—except that the shot from behind was fired from the Records Building, and was fired downward, and I am convinced it was with a sabot shell casing, by which you can fire a smaller caliber slug through a larger caliber rifle. It creates a little plastic jacket which keeps the gasses from escaping. I'm convinced they were doing this to implant a Mannlicher-Carcano slug in the body.

Now this hits him in the back, and actually appears to go through the back of the seat of the limousine, hits him right about here [indicates back], only goes in about as far as the second knuckle on your little finger.

Now Jack [JFK] has got this back brace. Jack suffers from what's known as an unstable back. He's in pain every day of his life. In fact, his back condition is so bad that I think all these sexual escapades attributed to him are beyond his physical capacity. He could not have been doing this. This is just exaggerated bullshit, intended to trash the man, so Americans no longer will care as much about his death…

So [JFK]'s hit in the back very near to the same time he's hit in the throat. This bullet goes through the windshield.

Everyone said the first shot sounded different; sounded like a firecracker. Okay, I got a guy, Jim Lewis down in the south, he's going to all these junkyards, he's firing high-velocity bullets through windshields. Guess what? They sound like a firecracker. So you got this round going through the windshield, hitting Jack.

Bob Livingston, world authority on the human brain [and scientific director of the National Institute for Mental Health and the National Institute for Neurological Diseases and Blindness in both the Eisenhower and Kennedy administrations], had gathered that the Secret Service had bought up a dozen of these windshields, they claimed for target practice. He believes it was to determine how much penetrating power you need to get through or to create a replacement windshield if you needed it. It turns out, you know, all of the above, because the limousine with the hole in the windshield—which is observed by two motorcycle patrolmen and a reporter for the *St. Louis Post-Dispatch*—

GONZO SCIENCE: It's in a photo too, isn't it?

FETZER: Yeah, it's even shown in the Altgens photograph. Also in frame 225 of the Zapruder film you can see it too; there's one frame there where it's very clear.

Plus, see, LBJ had the limo sent back to Ford on Monday, they had it stripped down to bare metal. Well, think about it, this is a crime scene on wheels, what are we doing? This belonged in the Smithsonian, unaltered.

In fact when they first got to Parkland [Hospital], Secret Service men got a bucket and sponge and started washing brains and blood out of the limousine. I wonder why they were doing that! We even have photographs of this. Jack's personal photographer, Cecil Staunton, took pictures of them washing blood and brains out of the limousine.

So, so far you have Jack hit twice. Once in the back—see, it may have even hit the brace, and this is why it's so shallow, about as deep as the second knuckle on your little finger, downward angle, between a 45- and a 60-degree angle—and now he's been shot in the throat and his hands are coming up to his throat.

Other shots now may have been taking place concurrently but miss, because one injures an innocent bystander. So if you try to break it into roughly three salvos, you can have one of these misses either hits the chrome strip over the windshield—and, you know, there are photographs of this—or it injures an innocent bystander, hitting the curbing.

So he [William Greer, the driver] continues to drive but pulls over to the left and brings the limousine to a halt. And we have 60 eyewitnesses who either report seeing it slow down dramatically or come to a halt, where, of course, in order to come to a halt, you have to slow down dramatically. The argument has been made, by these people who want to defend the Warren Report, that, "Well it's not consistent, some say it slowed, others say it actually stopped," ignoring the obvious explanation that it slowed as it came to a stop.

Now, Big John [Connally, Texas governor seated in front of JFK] is trying to figure out what's going on. He turns around to the right, he starts to turn around to the left. Boom, he feels, doubling up, a bullet has hit him right near his right armpit, shattered a rib, allegedly hitting him in the wrist, allegedly going on to impact in his thigh.

However, it looks more plausible that the shot to the wrist and the shot to the thigh may have been with separate bullets. The physician who was [responsible for] removing the metal from

Connally's wrist, for example—Robert Shaw by name—when told that the government was saying that the same bullet hit Jack and hit John, said he thought that that was most unlikely, since he had removed more lead from John Connally's wrist than was missing from this bullet. This of course is the alleged "magic bullet" that is virtually intact.

So bullets continue to be fired now, and one is going to hit Jack in the back of the head, and another's going to hit him in the right temple and these are in fairly close proximity after the limousine has been brought to a halt. And at least one more shot is hitting Connally. Another shot is being fired and winds up in the dirt.

So now you've got the shot that went into his back, you've got the shot that went into his throat, you've got one of the shots that missed, you've got the shot that hit Big John Connally, you've got a shot that hit Jack in the back of the head, you've got a shot that hit Jack in the right temple, you've got another shot that missed, you've got probably another one into Big John, but you've got another shot that went over to the—the one I just mentioned that missed and went into the dirt over there—but then you've got the shot that hit the bystander, and possibly you've got another that hit Big John, so you've got eight, nine, or ten shots fired from six different locations.

GONZO SCIENCE: In roughly three volleys, so it's kind of hard for the witnesses to tell just how many shots have been fired.

FETZER: In roughly three volleys, yeah. There are reverberations. It's hard to tell, depending on where you're standing, plus—people forget this—most of these weapons were using silencers. That was not a novelty.

GONZO SCIENCE: Sure. Because the shooters were in buildings—

FETZER: They were in buildings, they were scattered around here, there, you know.

I think it may be the rifle from the Dal-Tex building was the only unsilenced, because, see, that fired three shots, and it's in proximity to the Book Depository—it's the one physically closest—and the echo pattern might be such that it could easily convey the impression that it came from back there.

[The location in the Dal-Tex building that has been fingered by researchers as the Dal-Tex "sniper's nest" is] actually the closet of a uranium-mining operation that's a CIA front.

GONZO SCIENCE: Really.

FETZER: Yeah. I know. All so fucking blatant.

Now, the goddamn grassy knoll, behind it was this parking lot that was used by—guess what?—the Dallas police department. So the whole area is surrounded by government-controlled buildings or locations.

And [Secret Service agent/limo driver] Greer claimed that he hit the accelerator and took the limousine out of there after the second shot—by which he meant the shot that hit Big John. The government always insisted there were only three shots.

[Greer] claimed he hit the accelerator after the second shot that hit Big John. But he doesn't—you can even see in the Zapruder film—what we've got left—he sits there looking at Jack, staring right at him until his brains are blown out, and then, lickety-split, then and only then, does he pull this limousine out of Dealey Plaza. And he's already brought it to a halt, but the way the film has been edited, you can't tell that that's what's happening.

Then you see this violent back and to the left motion, back and to the left—which many have thought was the most important evidence of a conspiracy, and it is important evidence of conspiracy, but we also suspect it may in part have been an artifact of the way in which they edited the film—that they had a frame out of order so it wouldn't have been nearly so dramatic. And in fact, when the frames were published in the Warren Commission supporting volume, they had them in the order they were supposed to be, so somebody didn't quite get the word when they released the fucking thing.

GONZO SCIENCE: Didn't they transpose frames in *Life*, too?

FETZER: No. *Life* never published the frames that were revealing. *Life* published frames that were insignificant, apart from [frame] 313 where they showed the head spray, and said, "This is the shot that determined the direction from which…" As though a single point in time was sufficient to define a line, right? You need two points for a line. Any of us who have taken Euclidean geometry [would know that], right?

Plus you've got this spray spreading out much too fast. Plus then you have this big blob [in the Zapruder film during the head shot]. An expert in special effects from Hollywood, who received the Oscar in 2001, concluded the blob had been painted in.

GONZO SCIENCE: Really.

FETZER: Mm-hmm. So see, I put it on the cover [of my book *Murder in Dealey Plaza*]. I talk about the cover [in the book], and I say here, "The front cover illustration is a frame from the fake Zapruder film featuring 'the blob.'"

[The Zapruder family can] stuff it. They charge people thousands of dollars to reprint their frames; no one has ever used a frame of the Zapruder film on the cover of a book until I did. And I told them to stuff it. I didn't pay them a nickel. And they tried to approach me legally. I had a very competent attorney, and I was telling him exactly how I wanted to argue it because there are various provisions in the copyright law, for especially—not just fair use, but criticism. How can you criticize something if you can't reproduce it, because you've got to talk about what it is in order to criticize it? So there's a clear exception there.

Plus, a guy named Nimmer, who was the foremost expert on copyright law in the world until his demise, pointed out in one of his many volumes on the subject—there were two, he thought, clear examples of exceptions to copyright in the photographic realm. One was this famous picture of the Vietnamese girl running down the road after being napalm-bombed—remember her naked running down the street? And the other was the Zapruder film. . . .

The guy [the special effects expert from Hollywood] was looking at these two frames—frames 302 and 303—and he told Noel

Twyman that the reason why it's fuzzy in the background but sharp in the foreground in frame 302, but sharp in the background *and* in the foreground in frame 303, is because in 303 the limo is stationary and in 302 it's still moving. So you get the discrepancy between the motion in the foreground because you're panning to follow the limousine, see, so that's going to blur the background if it's stationary relative to the panning when you're moving your camera.

But these are supposed to be one-eighteenth of a second apart! Now, you can't go from in motion, you know, say 10 or 12 miles an hour, to stationary in an eighteenth of a second. But he also told Noel that he believed that "the blob" had been painted in. The blob moves around in a curious way...

There are more than 15 indications of Secret Service complicity in setting [JFK] up.

GONZO SCIENCE: Does this mean the Secret Service knew exactly what was going to happen, or were told to hang back that day?

FETZER: No, it was controlled. I mean, I'll tell you how bad it was. A fellow [in the Secret Service] named John Ready—how appropriate—started to respond in Dealey Plaza after shots began to be fired, and Emory Roberts, the guy in charge of the Secret Service detail, called him back. Called him back!

There was another Secret Service agent named Henry Rybka who, at Love Field [in Dallas before Dealey Plaza], started to run behind the limousine the way he'd been trained to do, and Emory Roberts called him off, too. And he's left behind at Love Field.

Plus, all the vehicles are in the wrong order. The presidential limousine is right up front. Where do you suppose the presidential limousine ought to be? It ought to be somewhere in the middle. So, you run the lowest dignitaries like the mayor first, then you run the vice president, then you get to the president, as the crowd cheers, and then they applaud. Hey, if you run Jack out there, then they cheer for Jack and then they leave, and the mayor doesn't have anyone to look at him.

Plus there was supposed to be a flatbed truck on which the press was going to ride. It normally would have preceded the presidential limousine so they could cover it with cameras and television. And it was cancelled.

There was a military aide to Jack Kennedy who normally rode in the front seat of the presidential limousine in between Greer, the driver, and [Roy] Kellerman, who was in charge of the detail. And they put him in the final vehicle, where he could be in the company of—guess who—Admiral George Berkeley, the president's personal physician. If he had ridden in the middle of the front seat, the shot through the windshield would have hit him, not Jack.

So the president's personal physician, who would be expected—you would want to keep him handy to render aid if you need him—is put in the last vehicle. Along with the press! They're all stashed in these back vehicles so they can't cover it and see what is going on.

GONZO SCIENCE: So no one can see.

FETZER: No one can see. A few photographers scattered around, mostly amateur people. One professional photographer, James Altgens, gets a famous photograph of looking at the Book Depository, seeing the Secret Service men looking around—the bullet's already gone through the windshield, Jack's already clutching at his throat, he's already been hit in the back. His brains haven't been blown out yet, though.

GONZO SCIENCE: And then a lot of people had their film confiscated at the scene.

FETZER: Oh, yes, yes. The important case being Beverly Oliver, who was a singer at a club near the Carousel Club, which of course was Jack Ruby's strip joint. And she had been standing right across the street from the grassy knoll, shooting toward the grassy knoll with a brand-new camera.

And within a couple of days, and FBI agent by the name of Regis Kennedy—who was on loan from New Orleans—gee, I wonder where Lee Oswald spent his days before Dallas—encounters her and asks for the film which she has not even developed. He takes it, assuring her it will be returned to her, but of course it disappears forever.

That probably would be the most important footage because it's right up toward where some of the shooters were. I mean, it really turns out—you know, who is going to be filming the left rear corner of Dealey Plaza where you have the triple underpass and the earth? And yet that appears to be in the location from which the shot through the windshield came.

I mean, Doug Weldon wrote a whole chapter on this [in *Murder in Dealey Plaza*], did a lot of research. He even went so far as to track down the employee at Ford Motor Company who was responsible for replacing the windshield that Monday, who of course confirmed the through-and-through hole.

But then the Secret Service presented another windshield, with little cracks caused by something from the inside, that they had indeed substituted. So Bob Livingston really was right on all these counts.

So [the Ford guy] put in a brand-new windshield, see, at Ford. But then the Secret Service fiddled around. They had several to work with and then they replaced it.

GONZO SCIENCE: So some Secret Service guys were central to the conspiracy?

FETZER: Setting up [JFK], oh yeah. The big players, I mean, you got the Secret Service setting him up, and there are more than 15 indications. And you got the CIA-Mafia, which had been engaged in taking out—attempts to take out Castro without success, but multiple attempts, and they had been part of a so-called Operation Mongoose, which Jack and Bobby, alas, had initiated after the aborted Bay of Pigs—and they placed Edward Lansdale in charge of [Mongoose], an Air Force general who became very famous worldwide. He pulled off assassinations all over the world.

He was photographed in Dealey Plaza after the assassination. And it is my belief, and that of others like Len Osanic, who was the aide to Fletcher Prouty [the person on whom Oliver Stone's "Colonel X" was based], that indeed this was an Ed Lansdale operation, that he actually supervised the execution of Jack Kennedy.

Now of course he would have been responsive to superior officers—if Lansdale were only one or two stars at the time—like three-star Charles Cabell, who was the deputy director of the CIA at the time of the Bay of Pigs, who called up Jack Kennedy at 4:30 in the morning in the company of Dean Rusk to insist that he provide the close air support that they thought he had promised, but which he declined to provide.

So when Cabell was dismissed along with Richard Bissell, the other deputy director, and subsequently Allen Dulles, who was director, Cabell returned to the Pentagon describing Jack Kennedy as a "traitor."

Now, it just happens that Charles Cabell was born in Dallas in 1903. His brother Earl, who was born near Dallas three years later, would become mayor of the city of Dallas in 1962—in which capacity he not only supervised the police department, but ceremonial activities including parade routes, where I'm quite convinced that this changed parade route, which is absolutely contrary to Secret Service policy, could never have been made without [Mayor Earl Cabell's] approval. This was the Cabells' back yard.

If you work your way up the Air Force chain of command you get Curtis LeMay. Now, he was a model for—do you remember the film *Dr. Strangelove*? The part played by Robert Ryan, this cigar-smoking colonel who was seeking to precipitate nuclear war? Well, that was based upon Curtis LeMay.

And of course if you work your way further up the chain of command, you find the chairman of the Joint Chiefs of Staff was a guy named Lyman Lemnitzer, a four-star Army general who, at the suggestion of President Eisenhower—who was very unhappy because Castro had come to power during his administration—suggested to the chiefs that if they didn't have a legitimate reason for attacking Cuba, they should invent one.

These guys took that very seriously, and they consulted—in their plans to come up with a scheme to take out Castro—Ed Lansdale. So Lansdale was involved in these schemes, what was called Operation Northwoods, and it's reported in a book called *Body of Secrets* written by James Bamford. It's about the National

Security Agency. He's also written an earlier book called *The Puzzle Palace* about the National Security Agency. But during his research he came up with all this stuff about Operation Northwoods.

So they came up with schemes such as blowing up the Atlas rocket that was going to carry astronaut John Glenn into space, and then blaming it on the Cubans; loading a commercial airliner full of college kids, flying it over Cuba, and shooting it down. They talk about how "the list of casualties in the papers would do wonders to inflame the population to rise up and crush Castro," right?

But Jack no longer trusted either the military or the CIA, and was having all these schemes vetted by Maxwell Taylor, and of course he was turning them down.

It didn't take that long for the chiefs to conclude that the real obstacle to taking decisive action against the communists, in Cuba and elsewhere around the world, was Jack Kennedy. So I think this brought them on board.

So that when the actual execution went down, the CIA-Mafia teams Lansdale had actually been employing in Operation Mongoose were probably used then to take out Jack Kennedy, with Ed Lansdale supervising, with the backing of the Joint Chiefs.

And then they covered it up...the FBI was actually using like a three-tiered filtration system: if a witness was in a good position and knew too much, you did not interview them; or if you mistakenly did interview them, then you didn't ask the crucial questions; and if you did interview them and asked the right questions, then you changed their testimony.

They had agents stationed in all the photo-processing plants in Dallas for two weeks after the assassination...Even Richard Trask in *Pictures of Pain* shows a photograph of the little card they left when they took your photographs, you know, "the government needs your stuff," right? So they were collecting this stuff immediately after the assassination.

Jean Hill, who was assisting her friend Mary Morman, took one of the most famous photographs immediately after Jack's head was whacked. You can actually see this patch of his hair on his shoulder, though many copies of this photograph are obscured. Jean was almost immediately apprehended and hauled off to some office in Dealey Plaza and interrogated for an hour and a half, and she said she'd heard at least four shots and that one of the shots had come from the grassy knoll. [She was told] that that couldn't be true, because there'd only been three shots, and if she knew what was good for her, she'd keep her mouth shut!

I mean, if the goddamn thing had just happened, how do you know there are only three shots? She's written a whole book about it called *JFK: The Last Dissenting Witness* with a guy named Bill Sloan.

GONZO SCIENCE: Who were the shooters? Do you have names?

FETZER: Well, they're multiple, multiple. Actually, interestingly, Johnny Rosselli claimed to have been one of them. He actually said he was one of them.

GONZO SCIENCE: Do you think he's credible?

FETZER: Oh, Rosselli's a pretty credible guy.

There are quite a few candidates [for shooters].

The shooter on top of the County Records building may have been a deputy sheriff by the name of Harry Weatherford.

The guy who was in the Dal-Tex building may have been a guy named Eugene Brading, also known as Jim Braden. He might also have been this guy Chuckie Nicoletti; Nicoletti may also have been in the Dal-Tex building.

I believe the shooter in the Book Depository may have been a Cuban [from eyewitness descriptions of a dark-skinned man in the window].

Why they were shooting [Texas Governor] Connally has not been obvious, but the fact that he was hit, you know, I don't see it as just a miss, in particular if he was hit two or three times. . . . But I now believe you are right to suggest to me that the person who was supposed to be sitting in that seat was Ralph Yarborough, LBJ's liberal political enemy, which explains not only why they were shooting at him but also why LBJ put up such a huge fuss over who should ride with JFK. When Jack insisted that the chief executive of the state should ride with the chief executive of the United States that morning, it must have been too late to change the arrangements that were in place. I am in debt to you, Jim, for this observation, which is simply brilliant!

Other shooters—one of the most distant, who fired through the windshield, the toughest shot—it looks like the best speculation would be Malcolm "Mac" Wallace. This was a very wonderful marksman who apparently had killed other people for Lyndon Johnson.

Madeline Duncan Brown, Lyndon's mistress, by whom he had a son, was a trap and game shooter. She used to go out to the gun club. She said that two weeks before the assassination she was out there and that Mac Wallace was there almost every day and that he never missed anything that he aimed at. He was really good! And her conjecture was that Mac Wallace was one of the shooters.

Charles Harrelson was there; Harrelson is serving a life sentence for the assassination of a federal judge with a high-powered rifle; he killed John "Maximum" Woods.

GONZO SCIENCE: Harrelson is [actor] Woody's dad, right?

FETZER: Yeah, Woody's dad. He was hired by a couple of drug dealers who were about to be sentenced and who didn't want to be sent up for life. And he actually said—

GONZO SCIENCE: He confessed.

FETZER: Yeah, he said that he killed Jack, by which I interpret he meant he fired the shot that entered the right temple.

Now, he was subsequently put on video, and he admitted that he said it, but said he was out of his mind at the time, and the very fact that he said it showed that he was out of his mind—isn't that a clever thing to say?…

I didn't tell you enough about Madeline Duncan Brown. See, she was at a social event the night before the assassination at the home of Clint Murchison, one of the wealthiest oilmen at the time. Also, H.L. Hunt was there, who advertised himself as the richest man in the world; he may well have been. Richard Nixon was there—

GONZO SCIENCE: This was in Dallas?

FETZER: Right outside Dallas. Huge ranch. Nixon had been driven there by a local Republican leader who happened to work in the same bank where Madeline was a young advertising executive.

J. Edgar Hoover was there. [Madeline] thought perhaps this was in honor of J. Edgar. Most people don't know Hoover used to visit Dallas all the time and hobnob with these rich oilmen who thought Dallas was the actual political center of the United States, given how much they were able to influence politics.

George Brown of Brown & Root, heavy construction world-wide—now a subsidiary of Halliburton, by the way—was there. After the Vietnam War went down, Brown & Root got a contract to dredge a new landing area in Kahm Rahn Bay, even though Vietnam has many magnificent natural ports. They were paid about a billion bucks to dredge a new one at Kahm Rahn Bay.

John J. McCloy was there [at Murchison's]. He was our former High Commissioner to Germany after World War II, you know? And he also had served as chief executive officer of Chase Manhattan Bank.

And late in the evening—although to Madeline it was quite unexpected—there were only a couple dozen people there and the party was starting to break up—when LBJ showed up. And he and these other heavy hitters disappeared into a conference room.

About 15 or 20 minutes later when they broke up, he strode over toward her, and she expected he was going to whisper sweet nothings in her ear; but instead he told her in a hateful tone of voice that after tomorrow, he wasn't going to have to put up with embarrassment from those Kennedy boys. The same words she says he repeated the following day, when he called her again and said the same thing. Evidently, he wanted her to get the message.

About six weeks later they had a rendezvous in the Driskill Hotel in Austin, during which she confronted him with rumors rampant in Dallas at the time that he had been involved since no one stood to gain more personally.

And Lyndon blew up at her, and told her that the oil boys and the CIA had decided that Jack had to be taken out.

I had more than a hundred conversations with Madeline Duncan Brown. I know her personally. I interviewed her at the Lancer National Conference. We have our interview on tape, and I believe what she has to tell us.

Plus, it's been corroborated by Billy Sol Estes, who was a wheeler-dealer involved in a lot of scandals down in Texas ripping off the government for private benefit, many of which involved LBJ, John Connally, and their cronies.

For example, Billy Sol was a mastermind behind a huge cotton allotment scandal, where they put up the cotton allotment as collateral for a huge loan from the Department of Agriculture. And when the department became suspicious, they sent an inspector down by the name of Henry Marshall. But Marshall couldn't find the cotton.

They could find Marshall, however, with five shots in him from a single bolt-action rifle—where of course, in typical Texas style, the Justice of the Peace declared the death a suicide. Now try it—try it! Try shooting yourself five times with a single shot bolt-action rifle! It's not easily done.

And Billy Sol gave an interview to a French investigative reporter named William Reymond that was published in a French magazine—whose title translates to *Monday, Tuesday, Wednesday*—in which he explains that Lyndon was in deep political trouble: he had the Bobby Baker scandal on the one hand, he had the Henry Marshall scandal on the other. There was also this TFX, this experimental fighter decision, where Boeing had produced a far superior plane that actually in fly-ups was flying circles around the General Dynamics plane, but General Dynamics was located in Texas, and General Dynamics was given a contract, right? So all these were swirling around him.

And [Billy Sol] told Reymond that Lyndon had his chief executive assistant, Cliff Carter, go down to Dallas to make sure all the arrangements were in place for the [JFK] assassination.

GONZO SCIENCE: So Lyndon was central—not just a beneficiary, but in on the planning?

FETZER: Yes, absolutely.

And of course Hoover was covering it up, so by the afternoon of the assassination he'd already concluded there were only three shots, lone assassin, blah blah blah.

And they alter the—they take the body—see, if you've got control of the body, I mean, you keep it under military control, send it to a military hospital, tell Jackie, "Well, Jack was a Navy man, so we really should send his body to Bethesda [Naval Hospital]."

Now you think about Bethesda. Not only was Jack a former Naval officer, that meant they had all his x-rays there, see? They had all the old x-rays and medical records. So if they're going to alter them and make substitutions, they can do that.

For example, I'm convinced the chest x-ray currently in the set now is an old chest x-ray. It is of Jack Kennedy, but they made a substitution so you wouldn't see the damage that was done by the bullet that hit him in the throat that fragmented up and down.

GONZO SCIENCE: Do you think Nixon was in the center of the conspiracy?

FETZER: They had to bring in Nixon; Nixon had nearly won the election. Jack had barely beat him. It was very likely that Nixon might become president of the United States in the near future. They didn't want somebody re-opening it so they brought him in early on.

Of course, he'd been the point man on the Bay of Pigs. It wasn't difficult to gain Nixon's assent to the assassination. And once they're implicated, then they're going to cover it up, and they got their hands dirty early on, right.

[This] Murchison meeting was a ratification meeting: "Do we go forward or not?" "Yeah, everything's okay, let's do it."

GONZO SCIENCE: Who were "the tramps"?

FETZER: Well, that's an interesting question. There were all kinds of strange people in Dealey Plaza that day. The first one appears to have been Charles Rogers, also known as Charles Montoya, also known as "the man on the grassy knoll." A book was written about him.

GONZO SCIENCE: And he was a psychotic killer?

FETZER: Well, [he] may have been; I mean, he appears to have murdered and dismembered his parents and put their parts in the refrigerator until he could flush them down the toilet, but the police got there before he was finished. And he just disappeared, though he seems to have turned up later.

The second, the tallest, [is] Charles Harrelson.

The third [is] a fellow named Chauncey Holt who worked as a contract agent for the CIA...I've got three or four pictures of me and Chauncey.

GONZO SCIENCE: And so what did Chauncey tell you?

FETZER: Chauncey was a counterfeiter who was working as a contract agent for the CIA. He had a long fascinating history with organized crime. At one point he was an accountant for Meyer Lansky, for example, who of course was the boss of bosses.

It was he who really brought the mobs together and created a national Cosa Nostra.

Chauncey did a lot in southern California, and eventually wound up—he was an artist, he was a shooter, he could make his own ammunition, he used to run a school for assassins.

The CIA would buy up all these proprietaries with innocuous-sounding names. Sometimes they'd change them. Well, in this case, it was the Los Angeles Stamp and Stationary Store—sounds pretty innocuous, about a five-story building, central L.A. The first three floors were a legitimate business, but then the CIA had the top two floors.

Among Chauncey's activities while he was there was preparing the forged documents for use by Lee Oswald when he was in New Orleans—the "Alek Heidell" identity documents were prepared there.

The first three floors, by the way, they had a legit business manufacturing—guess what? Police and sheriff and law enforcement badges—can't do better! Hey, if you need something, we'll do better than a phony badge, we'll give you a real one!

Chauncey was given instructions by his handler at the CIA, a fellow by the name of Philip Twombly, to prepare 15 sets of forged Secret Service credentials for use in and around Dealey Plaza.

And he prepared these 15 sets of forged Secret Service credentials and took them down there [to the Dealey Plaza area] but the red pickup truck, where he was supposed to leave them—in the vicinity of the police parking lot by the way—wasn't there. So he wandered around for a while, came back, and all of a sudden there it was. So he left them. He was headed back toward the railroad area—behind the police parking area—when he heard the shots.

He had been told there was going to be a non-violent incident to draw attention to our policy toward Cuba. Now think about this: "non-violent," "intended to draw attention to our policy in Cuba." How does it draw attention to our policy in Cuba? Well Oswald, when he was in New Orleans, was being "sheep dipped" as a pro-Castro communist sympathizer, right? Well if you have a pro-Castro communist sympathizer shooting at the president of the United States, that certainly invites attention to our policy in Cuba, doesn't it? I mean, believe me! Not non-violent, though, and I wonder how seriously Chauncey could have taken that "non-violent" aspect.

Chauncey had gone to a designated railway car, a boxcar, to make the getaway, and found himself in the company of Harrelson and Montoya. Actually he had provided them with weapons earlier when he got down there, with handguns, and he actually had a sophisticated radio himself. And he found himself in the railway car with them and he noticed that the railway car also was loaded with a lot of explosives and weapons.

GONZO SCIENCE: Really.

FETZER: Yes. Now the train started to pull out; they thought they were going to be okay, but then it was called back because a railroad switchman had observed something going on and had them come back. This time they opened the door and they apprehended them and they escorted them through Dealey Plaza.

And in one of the most famous photographs of the three tramps, you see the three of them walking along and there's a civilian walking between them and the fence of the building. Well that civilian is a rather distinctive guy; he has a characteristic ring, he has a characteristic walk, and he has been identified conclusively—even by Victor Krulak, commandant of the Marine Corps—as Edward Lansdale.

I got a guy [Mike Sparks] who runs a think tank that evaluates military equipment and advises the Pentagon who gave a talk at my invitation in Dallas, who was describing the assassination and why he thought it bore all the distinctive, signature trademarks of Ed Lansdale.

GONZO SCIENCE: So Lansdale's figurative fingerprints are all over the assassination?

FETZER: In my opinion, he's the guy who supervised the actual execution…

GONZO SCIENCE: Can you comment on the alleged Oswald doubles? How many were there?

FETZER: Well, it's very clear that there's somebody—even J. Edgar Hoover observed someone was impersonating Oswald down in Mexico City. They had the voice recordings—they weren't Oswald's voice; they had photographs—the photographs actually looked a great deal like a guy named Jim Hicks, who was photographed in Dealey Plaza and claimed to be a communication coordinator for the assassination.

But the very fact that you have somebody impersonating Oswald right off the bat implies a conspiracy. I mean, there's no way around it.

Here you had Hoover advising his different FBI agents-in-charge that someone had been impersonating Oswald down in Mexico City. That's really quite fascinating all by itself.

GONZO SCIENCE: And he didn't really look like Oswald, either.

FETZER: Oh, no, not at all. Then in Dallas you had people who were impersonating Oswald, going to have a telescopic sight mounted on a weapon; or going to test drive a car and driving at high speed and saying they were going to come into a lot of money but they like Russian vehicles better; then a guy out at the rifle range firing on somebody else's target and when he was told to stop doing it, he said he got carried away because he thought it was Kennedy—all this shit!

This is so phony and so superficial and so clearly implicating the conspiracy. I mean, Oswald didn't know how to drive, for example. And who is going to go out on the rifle range and start shooting on somebody else's target? I mean, it's obviously planted—you know, they could have been more subtle…

Oh, here's another possible shooter. Jack Lawrence was an expert shot in the Air Force, and he had come to work at the automobile agency that provided the vehicles for the motorcade. This motorcade is also very peculiar because all the cars are different makes and colors. Practically all official motorcades are uniformly black limousines, like that. This is a real mix, and I think for the obvious reason: then you could visually identify who's where.

GONZO SCIENCE: "Get the black limo."

FETZER: Right.

GONZO SCIENCE: What about Officer Tippit, the police officer also said to have been shot and killed by Oswald directly after the assassination of the president?

FETZER: Tippit's a real wild card. He certainly was not shot by Lee Oswald.

I mean, among other reasons, Tippit was shot with two different types of ammunition. And the first officer on the scene identified ammunition from an automatic, which of course automatically ejects, which makes sense. Oswald, however, only had a revolver. Can you imagine shooting a policeman four times—three times in the body and once in the head—interestingly, in the right temple—and then opening the chamber and pulling out the rounds so you can immediately leave them for the benefit of the police? This is so absurd as to be beyond imagining.

The officer there first initialed these automatic shell casings—and they're shorter casings than revolver casings and easy to distinguish, as Robert Groden, in *The Search for Lee Harvey Oswald*, explains—anyway, he initialed them. But they were subsequently replaced by revolver shell casings that were not initialed.

And the woman across the street by the name of Aquilla Clemons said two men had killed Tippit and neither of them looked like Oswald. And two men with two types of ammunition makes sense.

GONZO SCIENCE: And then why did they do it?

FETZER: I don't know precisely why they killed Tippit. Someone thought it was for body parts; that they also shipped his body back to Bethesda. Tippit was alleged to have borne a resemblance to Jack Kennedy. Shooting him in the right temple is curious, because that's where Jack, after all, suffered that wound.

GONZO SCIENCE: One of the more far-out conspiracy theories is that Tippit was actually in the limo posing as JFK.

FETZER: That sounds absurd.

GONZO SCIENCE: But his physical resemblance must be the genesis of that.

FETZER: There's been suspicion that Tippit could have been "Badgeman," a shooter on the grassy knoll wearing a police uniform. . . .

I mean you really had a lock on the legal system—as long as you could block out [Attorney General] Bobby [Kennedy]. I talk about that in *Assassination Science*; Katzenbach wrote this memorandum even the day they were burying the president in a formal state funeral. Katzenbach—the number two man at the Department of Justice—wrote this memorandum saying, "The world must be convinced that this was done by this lone assassin, he had no accomplices; we got to cut off speculation about his motives; the Russian defection, the Russian wife, all that—too pat; maybe a report from the FBI will do the job."

They really cut out Bobby. After LBJ became president they yanked the direct line out of Hoover's desk. Hoover hated it; Bobby had had a phone put in that he expected Hoover to answer personally; Hoover just despised it.

GONZO SCIENCE: What's Ruby's connection?

FETZER: Ruby was local arrangements chair for the mob. He had a much bigger role in arranging the assassination than he's ever been credited with…

Madeleine remembered I think a couple weeks before the assassination when Jack Ruby came over waving a piece of paper and said, "I bet you don't know what I got here." Madeleine said, "Okay Jack, what do you have?" And he said, "It's the parade route that son of a bitch is going to take when he comes to Dallas."

And others reported seeing Oswald with Ruby in the Carousel Club—some of whom disappeared, and some of whom have never been heard from since.

GONZO SCIENCE: JFK and RFK—they were afraid of a military *coup d'etat*, weren't they?

FETZER: Well, as well they should have been. Read *Body of Secrets*…Jack was worried about a military coup…

GONZO SCIENCE: Do you agree with George Michael Evica that Oswald may have thought he was penetrating—and foiling—an assassination plot?

FETZER: Oswald appears to have been recruited by the Office of Naval Intelligence when he was a Marine. If he had been studying Russian when he was in the Marine Corps, it was with his government's blessing. He probably was getting instruction from the Monterey School, where the Department of Defense runs a high-level rapid-instruction foreign language [program].

Russian's not an easy language to learn. Turns out he took a Russian language exam and scored 50 percent, and people say, "Well, he didn't know [Russian very well]"—well, he got 50 percent of the problems right! That's like giving somebody an exam in calculus and they only get 50 percent of the problems right. It's so difficult and so technical that if they get half of them right, it's really displaying a high level of competence, I mean relatively speaking. And they called him "Oswaldovich" and everything else. Well, I'm telling you as a former Marine Corps officer, we don't have any communists in the Marine Corps. I mean, that's so ludicrous it's unbelievable. You know everything there is to know about your men.

And then he was subsequently stationed at Atsugi, which was our most secure base in the entire American military arsenal. It was the source of the U-2 flights. And he was even a radar operator. He would have known the altitude at which these overflights were taking place.

Now, the Soviets knew they were going on, but they couldn't do anything about it. They were embarrassed because they didn't know the altitude so they couldn't shoot it down. And it looks as though his pseudo-defection, which took place on a Sunday when it had no legal significance, was really intended to bring him in so he could give information to the Soviets, I believe about the U-2.

GONZO SCIENCE: To embarrass Eisenhower [when the U-2 was brought down]?

FETZER: Absolutely. Because there's this forthcoming summit conference between Eisenhower and Khrushchev, and instead of Francis Gary Powers taking his cyanide tablet and doing the right thing by blowing up his plane, he allows it to crash, he parachutes down, they capture him, they're able to parade him around, you know, embarrassing Eisenhower, right, aborting the summit, and increasing cold war tensions instead of decreasing them.

Now who would be interested in increasing cold war tension? You tell me. Because they were clearly behind all of this.

So then [Oswald] works in Russia, he winds up dating the niece of a KGB official, an attractive woman, they wind up getting married. . . .

Apparently there were at least two other defectors to the Soviet Union who married Russian women and came back to the U.S.; you know, military defectors. One was in Chicago, maybe one down in Miami. In other words the CIA had its people planted around and could use them as assets when it needed them.

When Oswald came back, he wasn't debriefed; he wasn't treated as though he had been a traitor to the country; he wasn't prosecuted. In fact he was given money by a CIA front organization and relocates down in New Orleans. Then he gets his job at the O'Reilly Coffee Company, and he's being sheep dipped, and he's working with this right-winger—all the stuff portrayed in Oliver Stone's film *JFK*. . . .

After the assassination, the attorney general of Texas, Waggoner Carr, conducted his own investigation and discovered that Lee Oswald had been working as an informant for the FBI since 1962. They made him informant #179; he was getting paid $200 a month right up to the time of the assassination. And it looks as though the letter [Oswald] gave to [FBI agent] James Hosty may well have reflected that he had penetrated...a conspiracy to kill the president. He was a patsy.

He was photographed down in New Orleans—a lot of this was all staged—like he had this altercation with these anti-Castro Cubans because he's distributing pamphlets on a street corner; the altercation lasts around two or three minutes but it just happens to be captured by a television camera parked across the street. Now, you've got to be Johnny-on-the-spot, you've really got to be smart about your sources to be so well-positioned at such a time, right?

And they did this interview explaining he was a Marxist but not a communist—and I mean there's a theoretical basis, a philosophical basis for all that…In the backyard photographs, he's not only got his Mannlicher-Carcano, but the belt with the revolver he's supposed to kill Tippit with. And he's holding two communist newspapers.

Well, one's *The Worker* and one's *The Militant*. I've consulted with my friend Dick Hudelson, many years my colleague. He's now a professor over at UWS [University of Wisconsin System] who does stuff on the history of communism and all that. He tells me that those who followed *The Worker* and those who followed *The Militant* had such strong differences of view that when they met, they would get into fights and try to kill each other! So it's highly unlikely that someone would be subscribing to both.

Plus—and you're going to love this—it turns out that we know the dimensions of the newspapers so they constitute an internal ruler that you can use to tell the height of the person in the photograph, who is about a half a foot too short to be Lee Oswald.

Oswald has an extensive intelligence history. There's a book by Philip Melanson titled *Spy Saga* that's a record of this.

GONZO SCIENCE: I can't believe the irony of coming to talk to you about conspiracy theories and JFK, and then to have Wellstone drop out of the sky.

FETZER: I know. I mean, listen, I'm not taking anything for granted. There's a conservative professor of political science I like a lot, a newbie. He was walking out as I was coming in, and he [asked] if I'd heard. And I said, "I want to know how this happened," and he said, "You're kidding me." And I said, "No. I want to know. It's a hell of a coincidence." It's ridiculous. I gotta know! They can't even get to the plane yet; it's in such a remote area, and there are so many trees and bushes. I mean, I don't know how they already know everyone's dead; I just suppose there are no signs of life. But I'm extremely suspicious.

GONZO SCIENCE: Yeah, I'll be very interested to hear what the black box has to say.

FETZER: I predict: there's just not going to have been a black box on that plane.

(*As Fetzer predicted, there was no black box to be found amidst the wreckage of Wellstone's plane.*)

Recommended Reading: *Assassination Science, Murder in Dealey Plaza,* and *The Great Zapruder Film Hoax* edited by Jim Fetzer. He maintains a website devoted to JFK and related issues at www.assassinationscience.com.

POSTSCRIPTS:
Full Disclosure

I Was a Teenage UFO Contactee

By Jim Richardson

MY RELATIONSHIP WITH THE UFO BLOSSOMED EARLY. My elementary school library had a thorough selection of books about many paranormal topics and I was a strong reader. I became so interested in these topics that it got me a lot of attention. People in class seemed to spontaneously organize around me during library time to help me comb the shelves for books. It was like an unofficial research organization of fourth graders.

For a while during this time—at least a whole summer—I took the family camera outside every night and scanned the skies. My father became alarmed at what he perceived as my credulity. By way of teaching me a lesson, he faked a couple of easy UFO shots. They were pretty lousy but it was a good exercise to "debunk" his photos. I even faked one myself and showed it off to him, proud because I'd done a better job. His photos had too many visual clues in them about size and distance, so I made sure mine had no give-aways. But this episode did nothing to sway that early belief in alien visitation—nor does it ever in real-life ufology.

I performed an actual UFO investigation during those days. I had a pair of classmates who lived across a field from one another, and each claimed to have seen a UFO landing. I thought it could have been a sighting of the same object by these two separate witnesses. One of them, Kirsten (my first crush) drew me a picture—was it in crayon?—of the object she claimed to have seen. It was a domed disc with lights and a three-eyed monster looking out the porthole. My personal feelings for the witness swayed my judgment. I'm sure she liked the attention and was more or less innocently weaving a tall tale. I might not have actually believed her if my friend Bruce hadn't also claimed to have seen it too. Were they in cahoots? I'll never know. I took the bus over to Bruce's house after school one day, and he took me to the alleged landing site. We walked through the field to a roughly circular depression in the ground with kind of patchy grass in it. This is it, Bruce affirmed. This is where it landed.

Nothing came of it. No surefire conclusions were possible and so it lingered there in the past, an anomalous story to be chewed over, like most UFO reports. That ended my career as junior UFO investigator.

After the sixth grade, my family moved to a different town and a different school, and that's when my career as a teenage UFO contactee began.

I'm not sure exactly how it started, but I know I was overwhelmed with this new town and junior high school in general. I began to realize that I was kind of a dork compared to a lot of these other 13 year olds, and I started spinning fantastic stories in a drastic overcompensation for my wobbly self-esteem. I straight-facedly began telling everyone in class, including close friends and teachers, that I was regularly visited by aliens. I maintained that these aliens frequently took me to their planet, taught me their alphabet, and replaced me with a synthetic android while I was away so that I wouldn't be missed. I drew pictures of these aliens, made maps of their world, created the letters of their language, and offered all of it up as proof that I had in fact had contact with these space people. I even hatched a plan to use my chemistry set to create fake skin, which I planned to utilize in an attempt to further the ruse that, occasionally, I was my own android stand-in. (This plan never reached fruition.)

After a while I grew out of it. I stopped spinning these yarns around the time I kissed my first girl. Everyone mercifully forgot about it or let it drop. My interest in the UFO question and the larger question of the paranormal began to wane.

As an adult, the paranormal and the UFO mystery have reasserted themselves in my life, albeit in a different guise. The research that Allen and I have done have convinced us that the truth lies somewhere between the hard-core skeptics and the loosey-goosey believers.

What's more, instead of being embarrassed by my experience as a claimed contactee, I realize that I learned from it. I have brushed up against the human impulse to mythologize oneself as the center of a cosmic drama. When I hear about the famous con-tactees, like George Adamski or Claude Vorhilon-Rael, I feel a sense of kinship. It is also a sense of pity. These people are not liars as such. They merely got lost under the pressure of psychological forces that perhaps transcended them. The mistruths in these cases are less like deceit, and more like delusion.

I Was a Twentysomething Admirer of a New Age Weirdo

By Allen Richardson

I REMAIN CONVINCED THAT JOSE ARGUELLES' 1984 book *Earth Ascending: An Illustrated Treatise on the Law Governing Whole Systems* is a work of genius. Written at a time of supreme lucidity in his life, its effect on me was nothing short of life altering.

While on the tail end of a semester in Mexico, I wandered around Mexico City's magnificent Museum of Anthropology in a cynical daze. My fevered brain was trying hard to balance an affinity for the study of culture with an overwhelming conviction that anthropology, in its current form, had not moved far beyond its roots as a tool of pacification and expansion. Despite the awesome power of this dense concentration of Meso-American artifacts, I was moved only to think, "But they're all dead." Whoever these folks were, and whatever knowledge they might've had which could possibly influence the train wreck of late-twentieth century life, it didn't amount to a hill of beans. Dead as Dillinger. This was my mindset as I returned to the States.

I happened across a copy of *Earth Ascending* in a friend's apartment, and from the outset I could barely believe my eyes. Arguelles' transposition of text and "holonomic maps" seemed to be an advancement of my incubating ideas about a balanced format for the expression of ideas. A diagram in the introduction called "The horizon-zenith system with the four principal directions as the basis for orientation in the universe" also had an eerie correspondence to concepts I had been groping after.

Arguelles' model attempts to unify different spheres of human knowledge. In doing so he fleshes out the term "holonomic," coined by George Leonard in 1978's *The Silent Pulse*, used to define entities "in the nature of a hologram." According to Arguelles, "holonomic is a term descriptive of holistic knowing, i.e., knowing that is simultaneously intuitive and rational, scientific and artistic. Thus, holonomics describes the order of reality as well as the way we come to know and express that order . . . holonomy is the name of the 'new science': that which proceeds from wholes to parts and in which consciousness . . . play[s] a formative role in both the structure and evolution of the universe. . . . [Holonomics is] the law governing whole systems."

Arguelles goes on to quote philosopher Oliver Reiser, which was my introduction to one of the most gonzo notions of all time—the world brain. Reiser's book *This Holyest Earth* contains the following description of it: "the radiation belt (or psi-field), the two poles, between them generating the 'world sensorium'—the guiding field which controls the psychosocial evolution of mankind."

Arguelles overlays his system onto another extremely gonzo area, the alleged one-to-one correlation of the DNA code and the ancient Chinese oracle the I-Ching.

Arguelles's interpretation of the Mayan calendar serves as another primary component of his philosophy. Comprised of multiple overlapping calendrical and numerlogical systems overlaid on the complex mythology of the ancient Meso-American cultures, Arguelles focuses on the Tzolkin, a divinatory calendar. The Tzolkin is a perpetual calendar wherein the base 13 of Mayan numerology interlocks with 20 cosmological symbols to generate a 260-day cycle. This tool is still in use among the Highland Maya of Mexico and Guatemala, a surviving relic of the larger cycles in which it was embedded during the peak of Mayan civilization.

Arguelles claims that the Tzolkin is a sort of temporal fractal, or a tool to model history. In essence, time on all scales ultimately folds into or out of the Tzolkin, in which the Mayans' much vaunted astronomical skills are seen to correlate to various cosmic events. Arguelles interprets literally the so-called "end date" of the Mayan Long Count calendar as December 21, 2012. This is an evolutionary blast-off date. Arguelles also was the primary mover behind the "Harmonic Convergence," which was sort of the start of the countdown.

I was quite keen on all this for some time. I participated in a few study groups that made liberal use of terms like "galactic synchronization beam." It wasn't until I attempted to grok his follow-up work *The Mayan Factor* that it became clear that Jose was drifting further and further into the realm of prophecy. I began to backtrack and cross-reference his work with academic anthropology and archaeoastronomy. Certainly his interpretation of the Mayans as peaceful timekeepers was a relic of a time before the language was truly cracked by modern scholars. The Mayans were many things; peace loving was not one of them.

I still value his insights into the properties of time and resonance, and the solar wind's relationship to life on Earth. There are many books on the purported one-to-one correlation between DNA and the I-Ching.

Years after my interest had waned, a college friend sent me Jose's material from the intervening years. The numerology was thicker than ever, and Jose was obviously the only one qualified to interpret it. In one booklet, he finally cut loose: Arguelles proclaimed himself to be "Valum Votan," the third earthly incarnation of Mayan King Pacal Votan. Yikes.

Acknowledgments

Most of these essays and interviews originally appeared—in some form or another—as weekly installments in the alternative newspaper *The Ripsaw News* of Duluth, Minnesota. "Gonzo Science" quickly spread online at Ripsawnews.com and our own Gonzoscience.com. In addition, we were the house weirdos at Anomalist.com for a while, and Forteantimes.com and Paranormalnews.com have regularly linked to us. We would like to thank each of these venues for their support.

INDEX

P A R A V I E W

PARAVIEW
publishes quality works that focus on body, mind,
and spirit; the frontiers of science and culture;
and responsible business—areas related to
the transformation of society.

PARAVIEW PUBLISHING
offers books via three imprints.

PARAVIEW POCKET BOOKS
are traditionally published books co-published by
Paraview and Simon & Schuster's Pocket Books.

PARAVIEW PRESS and *PARAVIEW SPECIAL EDITIONS*
use digital print-on-demand technology to create
original paperbacks for niche audiences,
as well as reprints of previously
out-of-print titles.

For a complete list of **PARAVIEW** Publishing's books
and ordering information, please visit our website at
www.paraview.com, where you can also sign up
for our free monthly media guide.

TRANSFORMING THE WORLD
ONE BOOK AT A TIME

www.ingramcontent.com/pod-product-compliance
Lightning Source LLC
Chambersburg PA
CBHW031918190326
41519CB00007B/337